# REAL GASES

# ENGINEERING PHYSICS

## An International Series of Monographs

*Edited by*

ALI BULENT CAMBEL and ASCHER H. SHAPIRO

*Volume 1*. ALI BULENT CAMBEL, DONALD P. DUCLOS, and THOMAS P. ANDERSON. Real Gases. 1962

*Volume 2*. LEON BRILLOUIN. Tensors in Mechanics and Elasticity. *(In Preparation)*

# REAL GASES


**Ali Bulent Cambel**
*Gas Dynamics Laboratory*
*Department of Mechanical Engineering*
*Northwestern University*
*Evanston, Illinois*

**Donald P. Duclos**
*Plasma Propulsion Laboratory*
*Republic Aviation Corporation*
*Farmingdale, New York*

**Thomas P. Anderson**
*Gas Dynamics Laboratory*
*Department of Mechanical Engineering*
*Northwestern University*
*Evanston, Illinois*


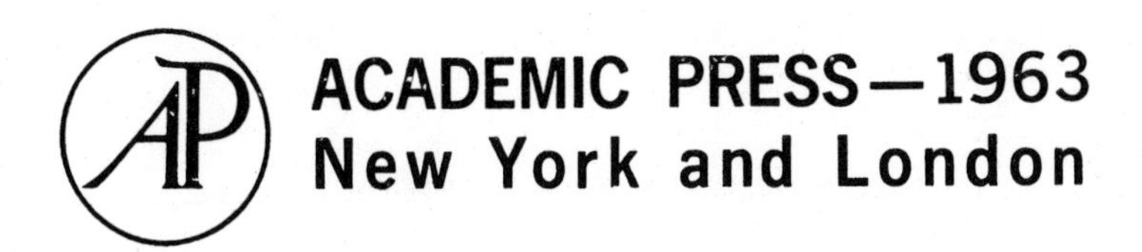

ACADEMIC PRESS—1963
New York and London

ACADEMIC PRESS INC.
111 Fifth Avenue, New York 3, New York

*United Kingdom Edition published by*
ACADEMIC PRESS INC. (LONDON) LTD.
Berkeley Square House, London W.1

*Library of Congress Catalog Card Number: 62-21935*

PRINTED IN THE UNITED STATES OF AMERICA

# Preface

This small monograph was prepared to serve as an introduction to the physics of real gases. In the context of this volume, a real gas is defined as one which is found at the high thermal energy levels commonly associated with aerospace type situations.

It should be mentioned that this volume is based on material covered in a number of research reports which were prepared by the authors in 1960, to evaluate the state of the art. The demand for these reports exceeded the supply and we were asked to consolidate them into one volume for the convenience of a wider audience. In doing so, major additions and revisions were made so as to include all pertinent information available through the middle of 1962.

We do not claim original contributions to the physics of real gases because the foundations have been laid previously elsewhere. However, we have attempted to interpret and evaluate how the theoretical foundations may be applied in the development of engineering devices. In essence then, our objective in preparing this volume was to review the art and the science underlying the behavior of real gas. We hope that in doing so, this monograph will serve as a bridge between equilibrium and nonequilibrium thermodynamics on the one side and high speed gas dynamics on the other side.

Although we have presented briefly the equations which govern the dynamics of real gases, we have excluded all detailed treatment of high speed gas dynamics because this may be found in several existing treatises.

As a rule the dissociated and ionized gases encountered in the aerospace technology are at relatively low pressures. Accordingly, the treatment presented in this volume is primarily applicable to high temperatures and low pressures. However, for the convenience of the reader we did include a very brief chapter on high pressure effects, because these may be found in some propulsive devices and laboratory research facilities.

The choice of a consistent set of symbols to designate the various parameters posed serious problems since this study covers aspects of gas dynamics, physics, thermodynamics, fluid mechanics, chemical kinetics, and aerodynamics. Although each discipline has a conventional nomenclature, they are not necessarily consistent with each other. In general, the symbols used in any section are those normally used in the field under discussion. For this reason, a separate List of Symbols has been compiled for each individual chapter and is located at the end of the chapter.

This work was done primarily under the sponsorship of the Department of the Air Force under Contract No. AF 40(600)-748 Arnold Engineering Development Center. The authors are indebted to Dr. B. H. Goethert, Director of Engineering, to Eino Latvala, Chief of Space Research and to W. McGregor of the Rocket Test Facility, all of ARO, Inc., operating company of AEDC. Without their support and continued stimulating discussions this work would have been impossible.

We are indebted to many of our co-workers, but particularly to Allen Fuhs, now of Aerospace Corporation; Serge Gratch, now at the Ford Motor Company; Richard Hoglund, now of Aeroneutronic, Inc.; and Stanley H. Jacobs, now at Harvard University; and Ching-Shi Liu and Arthur A. Kovitz of Northwestern University for a variety of contributions incorporated into this volume.

It is a pleasure to extend our appreciation to William Dorrance of Aerospace Corporation and Peter Wegener of Yale University. They perused portions of the manuscript and were generous with their time in pointing out shortcomings and recommending improvements.

Last but not least, we extend our appreciation to Elizabeth T. Anderson and to Patricia M. Duclos who with their wifely devotion assisted in the preparation of the manuscript.

ALI BULENT CAMBEL
DONALD P. DUCLOS
THOMAS P. ANDERSON

*Evanston, Illinois*
*Farmingdale, New York*
*November 1962*

# Contents

## Chapter 4

## The Debye-Hückel Theory for Ionized Gases



## Chapter 5

## High Pressure Real Gas Effects



## Chapter 6

## Dissociation and Recombination

## Chapter 7

## Ionization and Neutralization

*Chapter 1*

# Introduction to Real Gas Effects

## 1.1 Introduction

It is a well-known fact that when a vehicle travels at very high speeds, the air in its immediate vicinity is raised in temperature, and "real gas effects" come into play. Similar aerothermochemical phenomena must be considered in connection with the internal flow in propulsive devices. Because the behavior of real gases is appreciably different from that of a "perfect gas," engineering devices operating in the domain of real gases must be carefully designed. In this chapter some of the characteristics of real gases are discussed, as well as their behavior in engineering applications.

## 1.2 The Concept of a Real Gas

A gas may be characterized as being perfect or ideal according to different criteria. To the aeronautical engineer two considerations are of primary importance. Thus, one may speak of a gas which is perfect in a thermodynamic sense, or in turn one may speak of an ideal gas in a fluid mechanic sense. Qualitative descriptions of a gas in either category may be made. This will not be done in this monograph because excellent published treatises exist already (see for example refs. 1 and 2).

Here, a gas which is thermodynamically perfect is defined categorically by the thermal equation of state

$$p = \rho R T. \tag{1-1}$$

A gas is fluid mechanically ideal if it has no viscosity. Thus,

$$\mu = 0. \tag{1-2}$$

This monograph will concern itself particularly with the behavior of gases where Eq. (1–1) is not valid.

A real gas is defined simply as one for which

$$p \neq \rho RT.$$

One may then write for a real gas

$$p = Z\rho RT \tag{1-3}$$

where $Z$ is the departure (departure from perfect gas behavior) coefficient. Depending on the existing conditions, $Z \gtreqless 1$.

In a gas, real gas effects may be brought about primarily in any of three ways, namely, (1) by exposing the gas to very high temperatures, (2) by exposing the gas to very high pressures, or (3) by exposing the gas to high energy radiation.

The aeronautical engineer may encounter all of the above three modes. However, he will probably be most concerned with the effects brought about by high temperatures.

It should be noted parenthetically that at very low pressures a gas must be analyzed specially (see for example ref. 3). However, the regime of rarefied gas dynamics is outside the scope of this study.

A gas may assume a variety of degrees of freedom and with each there is associated an energy. Consider air at relatively low temperatures, say at considerably less than 1000 °K or corresponding to Mach numbers of $M < 2$ based on sea level conditions. One may represent a diatomic gas by a "dumbbell" or "rigid rotator" as depicted in Fig. 1–1.

At extremely low temperatures, namely, in the neighborhood of absolute zero, there will be three translational degrees of freedom, namely, in the $x$-, $y$-, and $z$-directions. At temperatures in the neighborhood of 10 °K, the rotational degrees of freedom are excited, although the molecular bond remains quite stiff. As the temperature is raised further, the vibrational degree of freedom comes into play. This generally occurs around 1000 °K. When the molecular bond is stretched to the breaking point, dissociation results. At atmospheric pressure, oxygen starts to dissociate at about 3000 °K, whereas nitrogen starts to dissociate at about 4500 °K.

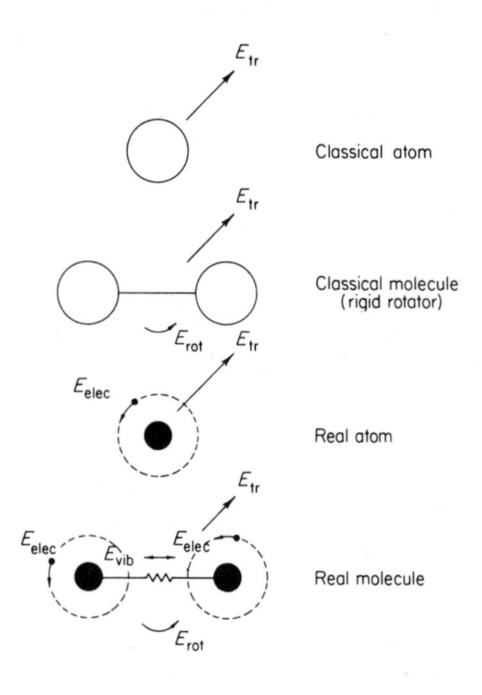


FIG. 1–1. Real gas models.

For example, for nitrogen dissociation we may write

$$N_2 + E_D \rightarrow 2N. \tag{1–4}$$

The energy of dissociation $E_D$ is commonly given in electron volts per molecule. The dissociation energy of nitrogen is 9.76 ev per molecule. At still higher temperatures, the electrons in the orbits around the atomic nucleus become excited to quantum states above the ground state. This excitation too contributes to the total energy, but the effect is negligible up to about 5000 °K. Finally, at still higher temperatures, electrons leave their orbits and the gas becomes ionized. For example, at about 10,000 °K (at atmospheric pressure) the ionization of the oxygen and the nitrogen in the air becomes sufficiently significant to be considered in engineering applications. For the single ionization of a nitrogen atom, one writes

$$\mathrm{N} + E_{\mathrm{I}} \rightarrow \mathrm{N}^{+} + e \tag{1-5}$$

where $E_{\mathrm{I}}$ is the ionization energy given in electron volts per electron.

If the number of free electrons and the number of positive ions are equal, the term plasma is also used, whenever the gas is ionized or electrically conducting. It should be noted that in the purest sense a plasma is electrically neutral, although electrically conducting. Because plasmas due to their electrical properties exhibit some markedly different characteristics, they have been called the "fourth state of matter," the other states being solid, liquid, and gas.

Both dissociation and ionization depend on pressure and temperature. Thus, a gas may be dissociated or ionized at a lower temperature if the pressure is reduced. For example, at atmospheric density the ionization of oxygen and nitrogen is significant at 10,000 °K, but ionization is significant at 6000 °K if the density is reduced to $10^{-6}$ times that at sea level.

In the preceding discussion the various levels of excitation were explained in a consecutive order. Actually this is not necessary, and in an actual application different levels of excitation occur not only in the aforementioned sequence but also concurrently. This is so because, during collisions, particles having higher energy (atoms, ions, and electrons) will impart their energy to the slower particles (such as molecules) exciting them, so to speak, prematurely.

The discussion so far made another simplification, namely, that one deals with a single component diatomic gas per se, such as oxygen alone or nitrogen alone. In the case of air, which is of course a mechanical mixture of many components, at least one other set of chemical reactions must be considered. This is the formation of nitric oxide (NO) as well as its subsequent dissociation and ionization. Thus, at atmosphere pressure and at about 1500 °K the formation of nitric oxide must be taken into consideration. In turn, at about 4000 °K the dissociation of nitric oxide becomes important at sea level pressures. At sea level the ionization of nitric oxide becomes appreciable at about 9000 °K (ref. 4). At low pressures nitric oxide dissociation and ionization become significant at considerably lower temperatures. Furthermore, it is believed that all charged particles which occur at temperatures below which nitrogen and oxygen ionize are due to the ionization of NO. Although the percentage of these particles is very small at these lower temperatures, their effect on some of the properties of air (such as the electrical conductivity) may be quite significant.

## 1.3 Real Gas Effects in Gas Dynamics

In Section 1.2 the behavior of gas particles in the real gas regime was outlined briefly. Indeed, it is necessary to understand the particle theory of gases if the properties of gases are to be described rigorously. On the other hand, the engineer charged with the design of propulsive devices and hypersonic vehicles must be familiar with the gross effects which may occur when a gas no longer behaves as a perfect gas. In this section an attempt is made to summarize the important aspects of real gas dynamics.

In chemically reacting media one must consider three major aspects of the problem. These are (1) thermodynamic aspects, (2) transport phenomena, and (3) fluid dynamics. Subjectively one might suggest that the thermodynamics and the transport properties associated with the flowing medium influence its fluid dynamics.

### 1.3.1 Thermodynamic Aspects

In Chapter 3 the thermodynamic considerations are treated in some detail. However, for the convenience of the reader, the following important thermodynamic differences between a real and a perfect gas should be mentioned.

As the temperature of a gas is increased, its specific heat rises markedly at first, but later may fluctuate. For example, at a moderately high temperature of 7000 °K, the isobaric specific heat of air is about 12 times that at sea level temperature.

The molecular weight decreases as the temperature increases.

The acoustic velocity is markedly influenced by real gas effects.

### 1.3.2 Transport Phenomena

The most important transport properties are thermal conductivity, diffusivity, and viscosity. In the ionized gas regime one must consider also the electrical conductivity.

In accounting for changes in the transport properties, one must recognize two flow regimes, namely, laminar flow and turbulent flow. It should be recalled that the transport properties are usually defined for the laminar conditions. However, most engineering applications are far from being laminar, and hence the convective and the eddy contributions must be included. This complicates the solution of problems because the eddy contributions depend on the flow and are not properties per se of the fluid. Thus, in establishing similarity (such as is commonly done by setting $\mathrm{Pr} = \mathrm{Sc} = 1$) one must proceed with caution.

If the dynamics of plasmas (with or without applied electromagnetic fields) is under consideration, one must recall that two regimes must be recognized further. These are where the short-range forces predominate (slight ionization) and where the long-range forces predominate (high ionization).

The case of short-range forces can be treated most adequately by kinetic theory and classical Boltzmann statistics. Although experimental data is lacking (due to difficulties in designing the necessary high temperature apparatus), the analytical aspects are well known. However, the case of long-range forces is not well understood analytically or experimentally. Here, one must consider multibody collisions, and the associated statistics become quite complex, requiring quantum mechanical techniques. It would not be an understatement to suggest that much work remains to be done in establishing an adequate theory for the transport phenomena in highly ionized plasmas.

The application of magnetic fields to ionized gases for propulsive as well as for flight control has been suggested, and much concentrated research is in progress. Here it must be pointed out that in magnetohydrodynamics the transport properties are no longer scalar quantities but become tensors. This too is an area which requires very extensive research before reasonable generalizations may be made.

### 1.3.3 Fluid Dynamics

The numerous aspects of real gas dynamics have been treated by Hayes and Probstein (ref. 5) and by Truitt (ref. 6). However, for the convenience of the reader a brief qualitative summary of real gas effects in aerodynamics is presented here. The treatment will seek to differentiate between a flow phenomena in perfect gases (see for example ref. 7) (constant specific heats and chemical composition) and in real gases (dissociated, excited, ionized). For details of each phenomena the reader should consult references given at the end of this chapter.

#### *1.3.3.1 Isentropic Expansion*

Where the initial conditions and final pressure are given, the effect of real gases on an isentropic expansion is to increase the final temperature and, thereby, the speed of sound and the velocity, and to decrease the final density. Because the effect on the speed of sound is greater than on the velocity, the Mach number is less for a real gas. The static temperature for a real gas is greater because the energy in the internal degrees of freedom† (which a perfect gas does not possess) is converted into translational energy during the expansion. There is consequently a pronounced real gas effect on area ratio, the real gas requiring a far greater area ratio to reach a given Mach number (ref. 8).

#### 1.3.3.2 *Prandtl-Meyer Expansion*

For given initial conditions and the same expansion angle, the effect of real gases is to increase the final pressure, temperature, and velocity, but to decrease the final Mach number (refs. 8 and 9).

#### 1.3.3.3 *Normal Shock Wave*

For a given shock velocity and free stream conditions, the effect of real gases is to increase the density ratio and the pressure ratio across the shock, while decreasing the temperature ratio (refs. 8 and 10).

#### 1.3.3.4 *Oblique Shock Wave*

For given initial conditions and wedge angle, the shock deflection angle is less for real gases than for perfect gases. Further, at any given Mach number, the limiting wedge angle (before a shock wave becomes detached) is greater for a real gas (ref. 1).

#### 1.3.3.5 *Shock Tube Performance*

For a given pressure ratio across the shock tube diaphragm, the effect of real gases is to decrease slightly the shock Mach number. However, if the shock Mach number is maintained the

† Translation is classified as an external degree of freedom, whereas rotation and vibration are internal degrees of freedom.

same, the stagnation temperature and the Reynolds number per foot are reduced for a real gas, while the flow Mach number behind the shock and the stagnation pressure are increased. The effect on the stagnation pressure is particularly large (refs. 1, 11, 12, and 13).

### *1.3.3.6 Boundary Layer Flow*

Real gases generally affect boundary layer flow in the following ways: the temperature in the boundary layer is decreased; the skin friction (except at high angles of attack) is decreased; the displacement thickness is decreased; the boundary layer induced pressure is decreased (ref. 14).

### *1.3.3.7 Magnetoaerodynamic Effects*

Where a magnetic field is applied to an ionized gas, the associated gas dynamics assumes many peculiarities which must be treated specially. This subject is outside the scope of this monograph and hence will not be treated here (ref. 15).

### *1.3.3.8 Heat Transfer*

In accounting for the heat transfer phenomena, one must differentiate between real gases without chemical reactions, real gas flow with chemical reactions, and magnetohydrodynamic heat transfer. When there is no chemical reaction, the heat transfer to the wall is less with real gases than with a perfect gas. This is explained by the fact that the peak temperature in the boundary layer and thus the temperature gradient at the wall is less with real gases. However, if chemical phenomena such as the recombination of dissociated gases occur, then appreciable energy is imparted to the wall, and hence heat transfer becomes a serious problem. The heat transfer in the stagnation regime of a blunt body may be reduced by magnetoaerodynamic arrangements (refs. 5, 6, 16, and 17).

## 1.4 Availability and Validity of Property Tables

In solving high temperature, gas dynamic problems, it is convenient to use tabulated properties. Fortunately there exists a variety of tables for equilibrium gas properties. Some of the most important of these tables will be discussed in this section.

In developing accurate tables for the equilibrium thermodynamic properties of air, two uncertainties were (a) the correct dissociation energy of molecular nitrogen and (b) the formation of NO. The complications with the nitrogen dissociation energy arose because the available spectroscopic data were consistent with two different models for nitrogen dissociation, one leading to a dissociation energy of 7.37 ev per molecule and the other to 9.76 ev per molecule. At first, the lower value was widely accepted as the correct one. In recent years, a number of experiments were performed which confirmed the higher value as the correct one. Most tables of thermodynamic properties prepared before 1955 use the incorrect dissociation energy of nitrogen, and consequently may be used only at temperatures below that for which nitrogen dissociation begins.

Over the years a large number of tables of the thermodynamic properties of air have been prepared. The report by Bethe (ref. 18) was one of the earliest. However, this report does not include the effects of NO formation and the calculation is based on a value of 7.37 ev for the dissociation energy of nitrogen. The widely used tables by Hirschfelder and Curtiss (ref. 19) and Krieger and White (ref. 20) were also based on the lower value of 7.37 ev, and were carried out for a rather limited range of pressure and density.

Hilsenrath *et al.* (ref. 21) computed the thermal properties of air, argon, carbon dioxide, carbon monoxide, hydrogen, nitrogen, oxygen, and steam for temperatures up to 3000 °K and for density ratios $(\rho/\rho_0)$ from 100 to $10^{-2}$. These tables are based on the correct value of the dissociation energy for nitrogen, but because of the uncertainty existing at the time the tables were being prepared, the range of temperatures was restricted to that below which dissociation is important.

Dommett (ref. 22) summarized the data on the thermodynamic properties of air which were reported up to early 1955. He also proposed values for the thermodynamic and transport properties of air in both dissociated and undissociated equilibrium for temperatures up to 12,000 °K. Because most of Dommett's references used the old value of the dissociation energy of nitrogen, the high temperature data in this report should not be used.

All succeeding tables of thermodynamic properties which are discussed in this section use the correct value for the nitrogen dissociation energy. It should also be noted that, in the tables to be discussed, air is assumed to be a mixture of perfect gases and, consequently, intermolecular forces are neglected.

In 1955, Gilmore (ref. 23) reported results for the composition, pressure, energy, and entropy of dry air at eleven temperatures between 1000° and 24,000 °K and eight densities between $10^{-6}$ and 10 times normal density. Gilmore assumed an ideal gas mixture in chemical equilibrium, including dissociation and ionization.

In 1956, Hilsenrath and Beckett (ref. 24) published tables of the thermodynamic properties of argon-free air for the temperature range 2000° – 15,000 °K. These tables represented an extension in range and improvement in presentation over the preliminary data reported in ref. 25. The properties tabulated are the compressibility factor, internal energy, enthalpy, entropy, and pressure. The properties are given at intervals of 200 °K between 2000° and 5000 °K, intervals of 500 °K between 5000° and 10,000 °K, and intervals of 1000 °K between 10,000° and 15,000 °K.

Moeckel and Weston (ref. 4) prepared charts showing the composition of air using Gilmore's (ref. 23) data, and the thermodynamic properties of air to 15,000 °K using Hilsenrath and Beckett's (ref. 24) data.

Logan (ref. 26) described procedures for the calculation of the equilibrium concentration and the thermodynamic properties of air in the temperature range 1000° – 20,000 °K.

In 1956, Treanor and Logan (ref. 27) presented tables of thermodynamic properties of air at high temperatures. The partition functions for the various components of high temperature air, temperature derivatives of partition functions, equilibrium constants for the high temperature air reactions, mole fraction concentrations of the constituents, and pressure, entropy, and enthalpy of the air were tabulated. These calculations were performed at 1000 °K intervals from 3000° to 10,000 °K and at eleven values of $\rho/\rho_0$ separated by factors of 10. The results of the calculations show good agreement with the calculations of Hilsenrath and Beckett and with Gilmore.

In 1957, Logan and Treanor (ref. 28) extended their previous report. The partition functions, first and second derivatives of partition functions, equilibrium constants, and temperature derivatives of equilibrium constants were tabulated at intervals of 100 °K between 3000° and 10,000 °K. The mole fraction composition, temperature and density derivatives of mole fraction composition, pressure, entropy, enthalpy, specific heats, and speed of sound were tabulated at the temperatures given above and at eleven values of $\rho/\rho_0$ between $10^{-6}$ and 100.

A more recent table of the thermodynamic properties of air is that of Hilsenrath, Klein, and Woolley (ref. 29, a revision of ref. 24). This, as well as the tables of Gilmore, Hilsenrath, and Beckett, and Logan and Treanor are all widely used today. The tables are of comparable accuracy; however, because of the different emphasis in each of these tables, the choice must depend on the particular problem to be solved. The range of temperatures covered as well as the temperature increments must be considered. Also, not all of the tables include particular items, such as partition functions, equilibrium constants, specific heats, and the speed of sound.

For specific applications to particular problems, specialized forms of the data given in the tables above have been developed. Perhaps the most widely known is Feldman's charts for traveling, reflected, and stationary shock waves (ref. 30). The thermodynamic properties are presented in graphical form as functions of shock speed and initial density. The charts were designed for conventional shock tubes and are limited to shock speeds less than approximately Mach 22. Ziemer (ref. 31) has extended Feldman's work to include the regimes found in electromagnetic shock tubes, $M = 50$. Other charts for shock phenomena in air are given in refs. 32, 33, and 34. For problems concerning hypersonic flow streams, Erickson and Creekmore (ref. 35) present air properties in the form of Mollier diagrams up to 4950 °R and Goin (ref. 36) gives Mach tables for stagnation temperatures up to 10,000 °K. For less specific applications, Hansen and Hodge (ref. 37) have tabulated air properties up to 15,000 °K as constant entropy properties of temperature.

Although the greatest bulk of the calculations of high temperature thermodynamic functions concerns air, since it is the most common gaseous medium, a moderate amount of data is available for other gases. Low temperature properties, up to 3000 °K, of argon, carbon dioxide, carbon monoxide, hydrogen, nitrogen, and steam are given in ref. 21 and for nitrogen in ref. 38. Thermodynamic properties of nitrogen up to 8000 °K and oxygen up to 5000 °K have been calculated by Treanor and Logan (refs. 39 and 40). Hydrogen properties are presented graphically by Gilmore (ref. 41) and King (ref. 42). Normal and reflected shock phenomena in hydrogen have been studied by Turner and the data are given in refs. 43 and 44.

A number of general studies on high temperature thermodynamic properties have been made and, although they do not present the thermodynamic data specifically as do the previously noted tables, they are quite useful at times. Woolley (ref. 45) discusses the general problem and Grabau (ref. 46) and Hansen and Heims (ref. 47) analyze air only. Some useful approximations are given by Hansen (ref. 48).

Tables and charts for the transport properties of high temperature gases over extended temperature ranges are not readily available due to the uncertainty associated with these calculations. Bauer and Zlotnik (ref. 49) have calculated transport coefficients for air to 8000 °K and estimate their error to be ± 25 to 30% even though this range is not considered to include "very high" temperatures. Transport properties for hydrogen are given by King (ref. 42).

## 1.5 The Speed of Sound

A thorough appreciation of the speed of sound is desirable in real gas dynamics for several reasons. Fluid mechanically, the speed of sound is used in defining important parameters such as the

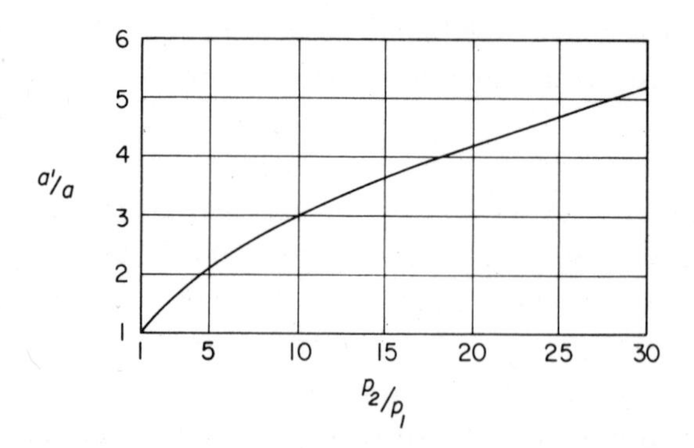


Fig. 1–2. Disturbance velocity ratio for air.

Mach number. Not only is the Mach number an indication of compressibility, but its magnitude also specifies whether the flow equations are elliptic, parabolic, or hyperbolic. Second, the acoustic velocity in a gas is a thermodynamic property of the gas from which other thermophysical properties may be calculated, and it is also an indication of relaxation phenomena as well as small irreversibilities.

By definition, the acoustic velocity is that velocity with which an infinitesimally small disturbance propagates isentropically through the medium. Thus, across a sound wave the state properties differ only a negligible amount. Figures 1–2 and 1–3 (from ref. 7) show how the disturbance velocity is affected by the

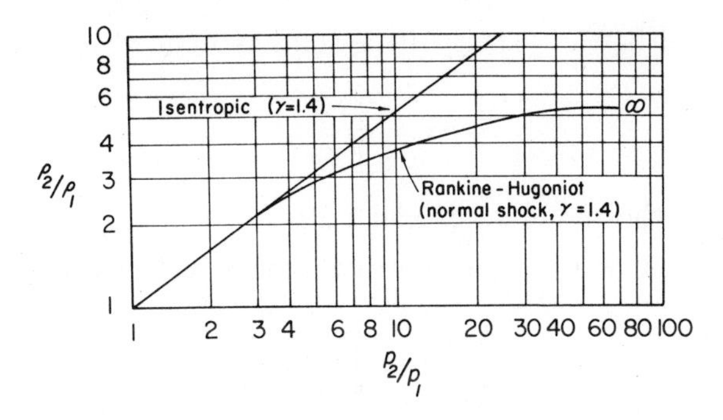


FIG. 1–3. Comparison of isentropic and Rankine-Hugoniot flows.

pressure ratio across the wave. Thus, in Fig. 1–2 $a$ is the sound velocity as defined, whereas $a'$ is the velocity with which any pressure disturbance such as a shock is propagated. It is evident that $a' \to a$ as $p_2 \to p_1$. In Fig. 1–3 the case of an adiabatic but not reversible shock is compared with the isentropic case. Again it is evident that the deviations occur as the properties change markedly. By definition then, the acoustic velocity is that velocity for which $p_2 \cong p_1$, $\rho_2 \cong \rho_1$, and $s_2 \cong s_1$.

It follows from the aforementioned remarks that large irreversibilities or discontinuities must be excluded from the acoustic velocity. However, there may be small irreversibilities due to transport phenomena or chemical reactions. These can and indeed must be considered in any discussion of the acoustic velocity in real gases.

The disturbance associated with a sound wave is due to alternate compressions and rarefactions which travel through a medium. The frequency with which these alternations occur depends on the type of disturbance which is occurring and bringing about the wave phenomenon. The frequency $\nu$ may be high or low, being specified as follows:

$$\nu = a/\lambda \tag{1-6}$$

where $\lambda$ is the wavelength and $\nu$ is the frequency.

The disturbance velocity in a solid is defined as

$$a' = \sqrt{\frac{E'}{\rho}} \tag{1-7}$$

where $E'$ is the modulus of elasticity.

In a liquid or in a gas the disturbance velocity is defined as

$$a' = \sqrt{\frac{G}{\rho}} \tag{1-8}$$

where $G$ is the adiabatic compression factor. A suitable equation for the acoustic velocity in a gas may be derived in a number of ways (see refs. 7, 13 and 50, for example) by combining the mass, momentum, and energy equations, and requiring further the existence of reversibility. A more rigorous, although less widely used approach, is due to Hirschfelder, Curtis, and Bird (ref. 51) and is an application of the entropy conservation equation. Thus, one arrives at the following equation:

$$a = \sqrt{\left(\frac{\partial p}{\partial \rho}\right)_s}\ . \tag{1-9}$$

It now remains to decide what functional relationship between the pressure $p$ and the density is to be introduced into Eq. (1–9). If the isentropic relation $p/\rho^\gamma =$ constant ($\gamma =$ constant) is used in Eq. (1–9), one arrives at the well-known expression for the acoustic velocity, namely,

$$a = \sqrt{\gamma RT}. \tag{1-10}$$

It should be noted that $\gamma$ is, by definition, the perfect gas specific heat ratio. However, if it is not correct to assume a perfect gas, or if there are deviations, the transition from Eq. (1–9) to Eq. (1–10) must proceed with extreme caution. Thus, to be considered are the dissipative effects which may exist or the real gas effects which may occur. These are diffusion, viscosity, thermal conductivity, electrical conductivity, chemical kinetics, and relaxation phenomena. Interpretation of these phenomena indicates that generally they may be classed under either of two categories, namely, the dissipations due to transport phenomena and the dissipations due to chemical phenomena.

When the dissipative terms are included in the acoustic velocity, the phenomena are irreversible, but this irreversibility is of the magnitude of the irreversibility encountered in the transport process associated with the dissipation. Thus, when dissipative terms are included, the process is not isentropic in the true sense. However, the deviation from equilibrium is small. Thus, Hirschfelder, Curtiss, and Bird (ref. 51) suggest that dissipations due to the transport phenomena may be neglected, but relaxation effects must be considered in evaluating the acoustic velocity.

When a gas in equilibrium is heated, its translational energy is increased. By virtue of the principle of equipartition of energy, the increase in this external mode of energy has to be distributed over two internal modes, namely, the rotational and the vibrational modes. The transfer of the energy affecting the external or translational mode takes place by virtue of collisions among the molecules. The transition of the energy into its rotational and vibrational modes requires a finite amount of time. As a matter of fact, the rotational energy reaches its equilibrium value more rapidly than does the vibrational energy. As in the process of heating, if a gas is suddenly cooled, its internal energy content is too great, and hence it becomes necessary to dissipate it externally by a large number of intermolecular collisions.

Any wave progressing through a medium provides a mechanism to add or remove energy locally with the gas because the passing of the wave brings about alternate compressions and rarefactions. If the frequency of these compressions and rarefactions is low, so that the period is large compared to the relaxation time in the gas, then an equilibrium sound velocity is defined. In other words, the gas always relaxes or reaches its equilibrium condition before the disturbance is passed through it. This acoustic velocity then specifies that the flow is not only isentropic but also experiences no change in the Gibbs free energy function.

$$a_e{}^2 = \left(\frac{\partial p}{\partial \rho}\right)_{\substack{s \\ \Delta F = 0}} \tag{1–11}$$

If, however, the period of the acoustic cycle is less than the relaxation time of the gas, then the gas may not reach its full equilibrium internal energy value. In this case one defines a frozen acoustic velocity,

$$a_f{}^2 = \left(\frac{\partial p}{\partial \rho}\right)_{\substack{s \\ [\beta] = 0}} \tag{1–12}$$

where $\beta$ accounts for the chemical effects. It should be noted in Eqs. (1–11) and (1–12) that both acoustic velocities assume the condition of isentropic behavior. However, in addition, they concern themselves with the Gibbs free energy and the chemical concentration. Because, $a_e \leqslant a_f$, one may write, after Gross and Eisen (ref. 52),

$$a_e{}^2 = a_f{}^2 + \sum_i \left(\frac{\partial p}{\partial \beta_i}\right)_{\substack{\rho \\ s \\ \Delta F = 0}} \left(\frac{\partial \beta_i}{\partial \rho}\right)_{\substack{s \\ \Delta F = 0}} \tag{1–13}$$

The dependence of the acoustic velocity on frequency is shown qualitatively in Fig. 1–4. When the chemical composition is fixed, the summation term in Eq. (1–13) disappears. Equation (1–13) may be used for dissociated as well as slightly ionized gases. Gross and Eisen (ref. 52) present the variations of $a_e$ and $a_f$ for hydrogen plasmas over wide ranges of temperature.

Springer and Gratch (ref. 53) have examined the speed with which an infinitesimal pressure disturbance is propagated through a reacting gas. They used a generalized rate law that is non-restrictive as to the kinetics of the reactions and applied the methods of Laplace transforms to obtain the propagation velocity. In this manner it was established that the wave front propagates at the frozen speed of sound except in the singular case of an infinite reaction rate.

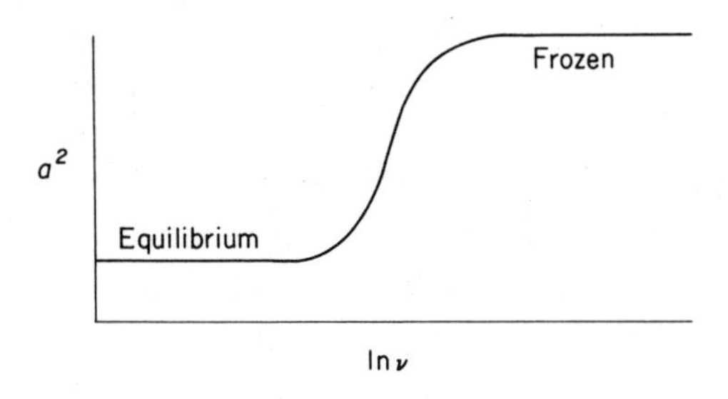


FIG. 1–4. Acoustic velocity dependence on frequency.

It follows from Eqs. (1–10)–(1–13) that one must differentiate between two "$\gamma$" terms. One is the usual specific heat ratio for a perfect gas and is constant. The other is an "effective $\gamma$." Thus, one defines the following $\gamma$-terms:

$$\gamma = \frac{C_p}{C_v}. \tag{1–14}$$

$$\gamma_{\text{eff}} = a^2 \frac{\rho}{p}. \tag{1–15}$$

Many gas dynamic equations incorporate a $\gamma$-term. Thus, in comparing calculations undertaken at different places, it is necessary that one be able to differentiate between the perfect gas specific heat ratio and the effective $\gamma_{\text{eff}}$ defined by Eqs. (1–14) and (1–15), respectively. In studying real gas phenomena, one should use Eq. (1–15). This may be quite simple if the necessary data for $a$, $p$, and $\rho$ are available. However, we can also compute $a$ from the specific heats because for a real gas which may be approximated by $p = Z\rho RT$, one has by definition

$$C_{v_{\text{eff}}} = \frac{R}{\gamma_{\text{eff}} - 1}. \tag{1-16}$$

Because the specific heats can be calculated for real gases with reasonable certainty, the $\gamma_{\text{eff}}$ term can be determined quite simply.

The acoustic velocity discussed so far is applicable to gases which may or may not experience chemical changes. We should note that in magneto-gas dynamics the applied magnetic field will have significant effects on the velocity of propagation of disturbances. The presence of a magnetic field creates an anisotropicity and thus several modes for propagation in which the propagation velocity is related to the velocity of sound. One such mode is the Alfvén mode, and two other modes result from the coupling of the acoustic waves and the Alfvén waves. A comprehensive discussion of linearized plasma oscillations is given by Oster (ref. 54).

## 1.6 *List of Symbols*

Latin symbols:

$a$ — acoustic velocity
$C_p$ — specific heat at constant pressure
$C_v$ — specific heat at constant volume
$E$ — energy
$E'$ — modulus of elasticity
$F$ — Gibbs free energy
$G$ — adiabatic compression factor
$M$ — Mach number
$p$ — absolute pressure
Pr — Prandtl number

$R$ — gas constant
$s$ — entropy
Sc — Schmidt number
$T$ — absolute temperature
$Z$ — departure coefficient

Greek symbols:
$\beta$ — concentration
$\gamma$ — ratio of specific heats
$\lambda$ — wavelength
$\mu$ — viscosity
$\nu$ — frequency
$\rho$ — density

Subscripts:
D — dissociation
e — equilibrium
eff — effective
f — frozen
I — ionization

## References

1. Rossini, F. D. (editor), *High Speed Aerodynamics and Jet Propulsion,* Vol. I: "Thermodynamics and Physics of Matter." Princeton Univ. Press, Princeton, New Jersey, 1955.
2. Condon, E. U., and H. Odishaw (editors), *Handbook of Physics.* McGraw-Hill, New York, 1958.
3. Patterson, G. N. *Molecular Flow of Gases.* John Wiley, New York, 1956.
4. Moeckel, W. E., and K. C. Weston, "Composition and Thermodynamic Properties of Air in Chemical Equilibrium." NACA TN 4265, April, 1958.
5. Hayes, W. D., and R. F. Probstein, *Hypersonic Flow Theory.* Academic Press, New York, 1959.
6. Truitt, R. W., *Hypersonic Aerodynamics.* Ronald Press, New York, 1959.
7. Cambel, A. B., and B. H. Jennings, *Gas Dynamics.* McGraw-Hill, New York, 1958.
8. Bird, G. A., "Some Methods of Evaluating Imperfect Gas Effects in Aerodynamic Studies." RAE TN 2488, January, 1957.

9. Heims, S. P., Prandtl-Meyer Expansion of Chemically Reacting Gases in Local Chemical and Thermodynamic Equilibrium." NACA TN 4230, March, 1958.

10. Wood, G. P., "Calculations of the Rate of Thermal Dissociation of Air Behind Normal Shock Waves at Mach Numbers of 10, 12, and 14." NACA TN 3634, April, 1956.

11. Squire, W., A. Hertzberg, and W. E. Smith, "Real Gas Effects in a Hypersonic Shock Tunnel." AEDC TN 55-14, March, 1955.

12. Stollery, J. L., "Real Gas Effects on Shock-Tube Performance at High Shock Strengths." RAE TN 2413, November, 1957.

13. Glass, I. I., and J. G. Hall, "Handbook of Supersonic Aerodynamics, Section 18, Shock Tubes." NAVORD Report 1488 (Vol. 6), December, 1959.

14. Burke, A. F., "Blunt Nose and Real Fluid Effects in Hypersonic Aerodynamics." IAS Paper No. 59-114, Presented at the IAS National Summer Meeting, Los Angeles, June 16-19, 1959.

15. Cambel, A. B., and J. B. Fenn (editors), *The Dynamics of Conducting Gases.* Northwestern Univ. Press, Evanston, Illinois, 1960.

16. Meyer, R. X., "On Reducing Aerodynamic Heat Transfer Rates by Magnetohydrodynamic Techniques." *Journal of the Aero/Space Sciences*, Vol. 25, No. 9, September, 1958, pp. 561-566.

17. Ferri, A., "Some Heat Transfer Problems in Hypersonic Flow." WADC TN 59-308, July, 1959.

18. Bethe, H. A., "The Specific Heat of Air up to 25,000°C." NDRC OSRD Report 369, February 9, 1942.

19. Hirschfelder, J. O., and C. F. Curtiss, "Thermodynamic Properties of Air, I and II." Univ. of Wisconsin, Reports CM-472 and CM-518, June 1, 1948, and December 21, 1948.

20. Krieger, F. J., and W. B. White, "The Composition and Thermodynamic Properties of Air at Temperatures from 500 to 8,000°K and Pressures from 0.00001 to 100 Atmospheres." The Rand Corp., Report R-149, April 15, 1949.

21. Hilsenrath, J., et al., "Tables of Thermal Properties of Gases." NBS Circular 564, November 1, 1955.

22. Dommett, R. L., "Thermodynamic Properties of Air at High Temperatures." RAE TN G.W. 429, August, 1956.

23. Gilmore. F. R., "Equilibrium Composition and Thermodynamic Properties of Air to 24,000°K." The Rand Corp., Report RM-1543, August 24, 1955.

24. Hilsenrath, J., and C. W. Beckett, "Tables of Thermodynamic Properties of Argon-Free Air to 15,000°K." AEDC TN 56-12, September, 1956.

25. Hilsenrath, J., and C. W. Beckett, "Thermodynamic Properties of Argon-Free Air." NBS Report 3991, April 1, 1955.

26. Logan, J. G. "The Calculation of the Thermodynamic Properties of Air at High Temperatures." Cornell Aeronautical Lab. Report AD-1052-A-1, May, 1958.

27. Treanor, C. E., and J. G. Logan, "Tables of Thermodynamic Properties of of Air from 3000°K to 10,000°K." Cornell Aeronautical Lab., Report AD–1052–A–2, June, 1956.
28. Logan, J. G., and C. E. Treanor, "Tables of Thermodynamic Properties of Air from 3000°K to 10,000°K." Cornell Aeronautical Lab., Report BE-1007-A-3, January, 1957.
29. Hilsenrath, J., M. Klein, and H. W. Woolley, "Tables of Thermodynamic Properties of Air Including Dissociation and Ionization from 1,500°K to 15,000 °K." AEDC TR 59–20, 1959.
30. Feldman, S., "Hypersonic Gas Dynamic Charts for Equilibrium Air." AVCO Research Lab., Research Report 40, January, 1957.
31. Ziemer, R. W., "Extended Hypervelocity Gas Dynamic Charts for Equilibrium Air." STL Physical Research Lab., Report STL/TR–60–0000–09093, April 14, 1960.
32. Hochstim, A. R., "Gas Properties Behind Normal Shocks at Hypersonic Velocities. I. Normal Shocks in Air." Convair Report ZPh(GP)–002, January 30, 1957.
33. Moeckel, W. E., "Oblique-Shock Relations at Hypersonic Speeds for Air in Chemical Equilibrium." NACA TN 3895, 1957.
34. Romig, M. F. "Conical Flow Parameters for Air in Dissociation Equilibrium." Convair Scientific Research Lab., Research Report 7, May 15, 1960.
35. Erickson, W. D., and H. S. Creekmore, "A Study of Equilibrium Real-Gas Effects in Hypersonic Air Nozzles, Including Charts of Thermodynamic Properties for Equilibrium Air." NASA TN D–231, April, 1960.
36. Goin, K. L., "Mach Tables for Real Gas Equilibrium Flow of Air in Hypervelocity Test Facilities with Total Temperatures to 10,000 °K." Sandia Corp. Monograph, SCR–288, March, 1961.
37. Hansen, C. F., and M. E. Hodge, "Constant Entropy Properties for an Approximate Model of Equilibrium Air." NASA TN D–352, January, 1961.
38. Woolley, H. W., "Thermodynamic Properties of Gaseous Nitrogen." NACA TN 3271, March, 1956.
39. Treanor, C. E., and J. G. Logan, "Thermodynamic Properties of Nitrogen from 2.000 °K to 8,000 °K." Cornell Aeronautical Lab., Report BE–1007–A–5, January, 1957.
40. Treanor, C. E., and J. G. Logan, "Thermodynamic Properties of Oxygen from 2,000° K to 5,000 °K." Cornell Aeronautical Lab., Report BE–1007–A–4, January, 1957.
41. Gilmore, F. R., "Approximate Thermodynamic Properties of Compressed Hydrogen Gas from 5,000 °K to 12,000 °K." The Rand Corp., Report RM–1650, March 7, 1956.
42. King, C. R., "Compilation of Thermodynamic Properties, Transport Properties, and Theoretical Rocket Performance of Gaseous Hydrogen." NASA TN D–275, April, 1960.

43. Turner, E. B., "Equilibrium Hydrodynamic Variables behind a Normal Shock Wave in Hydrogen." Space Technology Lab., GM TR 0165–00560, August, 1958.
44. Turner, E. B., "Equilibrium Hydrodynamic Variables behind a Reflected Shock Wave in Hydrogen." Space Technology Lab., STL/TR 59–0000–00744, June, 1959.
45. Woolley, H. W., "Effect of Dissociation on Thermodynamic Properties of Pure Diatomic Gases." NACA TN 3270, April, 1955.
46. Grabau, M., "A Method of Forming Continuous Empirical Equations for the Thermodynamic Properties of Air from Ambient Temperatures to 15,000 °K with Applications." AEDC TN 59–102, August, 1959.
47. Hansen, C. F., and S. P. Heims, "A Review of the Thermodynamic, Transport, and Chemical Reaction Rate Properties of High-Temperature Air." NACA TN 4359, July, 1958.
48. Hansen, C. F., "Approximation for the Thermodynamic and Transport Properties of High-Temperature Air." NASA TR R–50, 1959.
49. Bauer, E., and M. Zlotnik, "Transport Coefficients of Air to 8000 °K." *ARS Journal*, Vol. 29, No. 10, Part 1, October, 1959, pp. 721–728.
50. Liepmann, H., and A. Roshko, *Elements of Gasdynamics*. John Wiley, New York, 1957.
51. Hirschfelder, J. O., C. F. Curtiss, and R. B. Bird, *Molecular Theory of Gases and Liquids*. John Wiley, New York, 1954.
52. Gross, R. A., and C. L. Eisen, "Some Properties of a Hydrogen Plasma," in *Dynamics of Conducting Gases* (edited by A. B. Cambel and J. B. Fenn). Northwestern Univ. Press, Evanston, Illinois, 1960, pp. 15–24.
53. Springer, R. W., "The Speed of Sound in a Chemically Reacting Gas." M.S. Thesis, Northwestern University, April, 1962.
54. Oster, L. "Linearized Theory of Plasma Oscillations." *Reviews of Modern Physics*, Vol. 32, No. 1, January, 1960, pp. 141–168.

Chapter 2

# Introduction to Aerothermochemical Analysis

## 2.1 Introduction

In the previous chapter the behavior of real gases was outlined in a qualitative sense. It was shown that so-called real gas effects must be considered if (a) the higher energy levels (such as the vibrational mode) are excited; (b) if the gas is dissociating; and (c) if the gas is being ionized. A gas which is real in the sense of (b) or (c) is said to be chemically reacting, although the term physicochemically reacting might have been better. In a reacting gas one must consider not only the conservation of macroscopic mass, but also the conservation of the species or components involved in the reaction. In this chapter the analysis of the compressible flow of a reacting medium will be outlined.

In solving a real gas dynamic problem, one must (a) satisfy a number of equations of change; (b) specify the correct equations of state; and (c) define a number of dimensionless parameters. The following atlas outlines the various items which must be known in a gas dynamic problem.

Atlas of Gas Dynamic Analysis

A. Equations of change
   1. Mass
   2. Momentum
   3. Energy

B. Equations of state
   1. Thermal
   2. Caloric

C. Thermodynamic and transport properties
   1. Thermodynamic properties
   2. Transport properties
   3. Relaxation phenomena
   4. Acoustic velocity

D. Dimensionless parameters
   1. Thermal effects
   2. Dynamic effects
   3. Kinematic effects
   4. Inertia effects
   5. Chemical effects

## 2.2 Equations of Change

Two types of equations of change (conservation equations) must be recognized, namely, the *species* or component equations and the *global* equations. Although one may write the species equations for the conservation of mass, momentum, and energy, most problems are such that this is not necessary. Thus, one commonly writes the continuity equation in both the species and the global forms, whereas the momentum and energy equations are written in global form only.

### 2.2.1 Species Continuity Equation

The species continuity equation accounts for the formation or depletion of reactants in a reacting medium. It is written

$$\frac{\partial n_i}{\partial t} + \nabla \cdot (n_i \mathbf{v}_i) = \dot{\omega}_i. \tag{2–1}$$

Because Eq. (2–1) is written for $i$ components, it follows that there will be $i$ equations. Consider, for example, the partial but single ionization of argon, namely,

$$\mathrm{A}^0 = \mathrm{A}^+ + e.$$

Because the mixture will contain $A^0$, $A^+$, and $e$, Eq. (2–1) is written as follows:

$$\frac{\partial n_{A^0}}{\partial t} + \nabla \cdot (n_{A^0} \mathbf{v}_{A^0}) = \dot{\omega}_{A}, \qquad (2\text{–}2)$$

$$\frac{\partial n_{A^+}}{\partial t} + \nabla \cdot (n_{A^+} \mathbf{v}_{A^+}) = \dot{\omega}_{A^+}, \qquad (2\text{–}3)$$

$$\frac{\partial n_e}{\partial t} + \nabla \cdot (n_e \mathbf{v}_e) = \dot{\omega}_e. \qquad (2\text{–}4)$$

Because, by definition, in a singly ionized argon plasma $n_{A^+} = n_e$, the set of Eqs. (2–2)–(2–4) becomes simpler. However, it should be noted that the ion velocity is not necessarily the same as the electron velocity.

At the present the principle difficulty associated with Eq. (2–1) is the extreme paucity of accurate data for the reaction rate $\dot{\omega}$.

### 2.2.2 Global Continuity Equation

The global continuity equation is written

$$\frac{\partial \rho}{\partial t} + \nabla \cdot (\rho \mathbf{V}) = 0 \qquad (2\text{–}5)$$

where $\mathbf{V}$ is the average macroscopic stream velocity with respect to a fixed observer.

One may arrive at Eq. (2–5) from Eq. (2–1). In doing this, one recalls the following relations:

$$\rho = \sum_i n_i M_i, \qquad \text{(a)}$$

$$\mathbf{v}_i = \mathbf{V}_i + \mathbf{V}, \qquad \text{(b)}$$

$$\sum_i n_i M_i \mathbf{V}_i = 0, \qquad \text{(c)}$$

$$\sum_i M_i \dot{\omega}_i = 0. \qquad \text{(d)}$$

Substituting relations (a), (b), (c), and (d) in Eq. (2–1), multiplying by $M_i$, and summing over all components $i$ yields Eq. (2–5) which is the commonly encountered equation expressing conservation of mass.

### 2.2.3 Momentum Equation

Although the momentum equation may be written on both the species and the global bases, the latter form is usually preferred. This is because the global momentum equation is by far the more meaningful from an engineering viewpoint as long as the long-range forces among the particles need not be considered. However, in highly ionized plasmas the global equation may not be sufficient, particularly if magneto-gas dynamic effects must be considered.

The global momentum equation is written as follows:

$$\rho \frac{d\mathbf{V}}{dt} = \rho \mathbf{F} - \nabla p + (\nabla \cdot \mu \nabla)\mathbf{V} + \tfrac{1}{3}\nabla(\mu \nabla \cdot \mathbf{V}). \qquad (2\text{–}6)$$

If the problem under consideration is one in magneto-gas dynamics, then the magnetic momentum term $\mathbf{J} \times \mathbf{B}$ must be added to the right-hand side in Eq. (2–6).

### 2.2.4 Energy Equation

The energy equation is commonly written in global form and is

$$\rho \frac{d}{dt}(C_v T) = \nabla \cdot (k \nabla T) - p(\nabla \cdot \mathbf{V}) + \Phi \qquad (2\text{–}7)$$

where the dissipation function $\Phi$ is defined as

$$\Phi = \mu \left\{ 2\left[\left(\frac{\partial u}{\partial x}\right)^2 + \left(\frac{\partial v}{\partial y}\right)^2 + \left(\frac{\partial w}{\partial z}\right)^2\right] + \left(\frac{\partial w}{\partial y} + \frac{\partial v}{\partial z}\right)^2 + \left(\frac{\partial u}{\partial z} + \frac{\partial w}{\partial x}\right)^2 \right.$$
$$\left. + \left(\frac{\partial v}{\partial x} + \frac{\partial u}{\partial y}\right)^2 - \frac{2}{3}\left(\frac{\partial u}{\partial x} + \frac{\partial v}{\partial y} + \frac{\partial w}{\partial z}\right) \right\}.$$

If the problem is one in magneto-gas dynamics, the Ohmic heating must be included in Eq. (2–7) by introducing $J^2/\sigma$ on the right-hand side. Further if radiation is appreciable, then the radiation term $Q_R$ must be added.

## 2.3 Equations of State

In the preceding section the conservation equations for species, mass, momentum, and energy were outlined. The dynamics of a reacting gas are described completely only if these equations can be satisfied. To do this, it is necessary to know the thermodynamic properties of the medium. These are defined by appropriate equations of state (as opposed to equations of change already mentioned).

Two types of equations of state must be recognized, namely, thermal equations of state and caloric equations of state. These are discussed in greater detail in Chapter 3. However, a brief introduction is included here for the convenience of the reader.

### 2.3.1 Thermal Equations of State

Ordinarily, the thermodynamic state of a system is located on a surface in three-dimensional space. One then writes a functional relation between the pressure, density, and temperature, namely,

$$\phi(p, \rho, T) = 0. \tag{2–8}$$

The form of the functional relation must be determined experimentally or by analytic deductions of the assumed model. For a perfect gas one may write,

$$p = \rho RT. \tag{2–9}$$

For a reacting gas, Eqs. (2–8) and (2–9) are inadequate because they do not account for the chemistry of the problem. Even for a nonreacting real gas, Eq. (2–9) is not valid, and must be adjusted by virtue of the departure coefficient $Z$:

$$Z = \frac{p}{\rho RT}. \tag{2–10}$$

For a reacting gas, Eq. (2–8) should be written to account for the chemistry. Thus,

$$\phi(p, \rho, T, \beta_i) = 0 \tag{2–11}$$

where $\beta_i$ is the number of moles of the $i$th component.

### 2.3.2 Caloric Equations of State

Caloric equations of state define the properties of a gas, whereas thermal equations define the state. The caloric equations of state must account for the energy terms in the gas.

Consider the enthalpy for example. Thus, one may write

$$h = h(p, T) \tag{2-12}$$

if the gas is nonreacting. However, if it undergoes a chemical reaction, one should write

$$h = h(p, T, \beta_i) \tag{2-13}$$

where $\beta_i$ accounts for the chemical effects.

## 2.4 Thermodynamic and Transport Properties

In seeking a qualitative answer to a problem in real gas dynamics, it is necessary to know the thermodynamic properties such as the specific heats and the acoustic velocity as well as the transport properties such as diffusivity, viscosity, and conductivity. The properties as such are being determined and revised constantly, and it would be well to contact two important centers for such research, namely, the National Bureau of Standards and the Thermophysical Properties Research Center at Purdue University. A survey of available tables of thermodynamic functions is given in Section 1.4.

## 2.5 Dimensionless Parameters

The properties of a gas as well as the behavior of gases under various flow conditions may be defined and/or calculated from pertinent dimensionless parameters. Such parameters may also prove to be useful in establishing similarity criteria. The most important are the Reynolds number, the Mach number, the Prandtl number, the Schmidt number, the specific heat ratio, and Damköhler's first ratio. Additional parameters must be defined in magneto-gas dynamics such as the Hartmann number, the magnetic Reynolds number, etc.

In studying the dynamics of real gases, probably the most important additional parameter is Damköhler's first ratio, Dam–1. This is defined

$$\text{Dam–1} = \frac{t_{\text{trans}}}{t_{\text{relax}}} \tag{2–14}$$

where $t_{\text{trans}}$ is the translational time and $t_{\text{relax}}$ is the relaxation time. The former will be obtained from the fluid mechanics, whereas the relaxation time is determined from the thermodynamics and the kinetics of the gas. Two important cases must be recognized. Thus, when $\text{Dam} > 1$, equilibrium flow exists, whereas when $\text{Dam} < 1$, frozen flow exists. In the computation of Dam — 1, the main problem is the relaxation time, now not well known at all.

## 2.6 List of Symbols

Latin symbols:

$\mathbf{B}$ — magnetic field intensity
$C_v$ — specific heat at constant volume
Dam–1 — Damköhler's first ratio
$\mathbf{F}$ — body force
$h$ — enthalpy
$\mathbf{J}$ — current
$k$ — thermal conductivity
$M$ — molecular weight
$n$ — molal concentration
$p$ — absolute pressure

$t$ — time
$T$ — absolute temperature
$u, v, w$ — velocity components in $x$-, $y$-, and $z$-directions, respectively
$\mathbf{v}$ — average species velocity
$\mathbf{V}$ — average flow velocity
$\mathbf{V}_i$ — diffusion velocity
$x, y, z$ — Cartesian coordinates

Greek symbols:

$\beta$ — chemical concentration
$\mu$ — dynamic viscosity
$\rho$ — density
$\sigma$ — conductivity
$\Phi$ — dissipation function
$\dot{\omega}$ — rate of production or depletion of species

Subscripts:

$i$ — species $i$
relax — relaxation
trans — translational

*Chapter 3*

# Thermal and Caloric Equations of State

## 3.1 Introduction

Thermodynamic properties and states of a gas are defined by appropriate equations of state. Classical thermodynamics gives relationships between properties, but cannot predict the actual magnitude of these properties. For low temperatures, classical concepts can be augmented by experimental data and the properties and states readily determined. However, the generation of high temperature gases and the measurement of their properties is difficult, and in many cases impossible within the present state of knowledge of direct experiments. Therefore, the thermodynamic functions of high temperature gases often must be predicted theoretically.

The effects of the various modes of energy on the thermodynamic functions can be predicted by means of statistical thermodynamics, a branch of statistical mechanics. In this chapter statistical thermodynamic formulas will be given from which the thermodynamic functions can be evaluated. It should be emphasized that most of these formulas apply only to a perfect gas in thermodynamic equilibrium. The detailed mathematical derivations of these equations will not be discussed, and the interested reader is referred to any of the many excellent texts on statistical mechanics for these details.

## 3.2 The Partition Function

By means of Maxwell-Boltzmann statistics, it is found that the number of particles $N_i$ in the $i$th cell of phase space in the state of maximum thermodynamic probability, that is, the number of particles having a total energy in the range $E_i$ to $E_i + \delta E_i$, is

$$N_i = \frac{g_i}{B} \exp\left(-\frac{E_i}{kT}\right) \tag{3-1}$$

where $k$ is Boltzmann's constant, $T$ is the temperature, $g_i$ is called the degeneracy or statistical weight, and $B$ is a quantity to be defined. The sum of all the $N_i$'s is equal to the total number of particles, $N$:

$$N = \sum_i N_i = \frac{1}{B} \sum_i g_i \exp\left(-\frac{E_i}{kT}\right). \tag{3-2}$$

Therefore

$$B = \frac{1}{N} \sum_i g_i \exp\left(-\frac{E_i}{kT}\right). \tag{3-3}$$

Letting

$$Z = \sum_i g_i \exp\left(-\frac{E_i}{kT}\right) \tag{3-4}$$

we obtain

$$N_i = \frac{N}{Z} g_i \exp\left(-\frac{E_i}{kT}\right). \tag{3-5}$$

The quantity $Z$ plays an important role in statistical theory, and is called the partition function. The use of the partition function permits the derivation of expressions for all thermodynamic properties of the system. These expressions are applicable to molecules of all types. Only one condition need be made, namely, that the system consist of noninteracting or at least, weakly interacting particles. This means that the system must consist of an ideal, or virtually ideal gas.

When the partition function is written as in Eq. (3–4), the summation is over the distinct energy states only. The degeneracy $g_i$ is an integer denoting the multiplicity of the energy state $E_i$, that is, the number of states which have the same energy $E_i$. The partition function can be expressed therefore in the alternate form

$$Z = \sum_j \exp\left(-\frac{E_j}{kT}\right) \tag{3-6}$$

where the summation is over all of the energy states.

It is found that the energy of a molecule $E$ can be expressed as the sum of the translational part and an internal part, or

$$E = E^{\text{tr}} + E^{\text{int}}. \tag{3-7}$$

($E^{\text{int}}$ should not be confused with the thermodynamic internal energy of a system.) To a fair degree of approximation, the internal part of the energy is assumed to be the sum of three independent parts, namely, the rotational, vibrational, and electronic energies, or

$$E^{\text{int}} = E^{\text{rot}} + E^{\text{vib}} + E^{\text{elec}} \tag{3-8}$$

Therefore, the partition function becomes

$$Z = \sum_{jk} g_j^{\text{tr}} g_k^{\text{int}} \exp\left[-(E_j^{\text{tr}} + E_k^{\text{int}})/kT\right] \tag{3-9}$$

which can be shown to be equivalent to

$$\begin{aligned} Z &= Z^{\text{tr}} Z^{\text{int}} \\ &= Z^{\text{tr}} Z^{\text{rot}} Z^{\text{vib}} Z^{\text{elec}}. \end{aligned} \tag{3-10}$$

Alternately, assuming that the partition function could be written in the form of Eq. (3–10), thus assuming that the various energy modes do not interact with each other, would result in the condition that the gas being considered obeys the perfect gas equation of state.

## 3.3 Thermodynamic Properties

As stated previously, all of the thermodynamic functions of a system can be expressed in terms of the partition function for the particles making up the gas or its derivatives. The partition function, as defined by Eq. (3–6), has been derived for a single particle or a system of independent localized particles such as a crystal. The partition function for $N$ gas particles, which is a system of nonlocalized particles, is given by

$$\frac{Z^N}{N!}.$$

The factorial is necessary because in the derivation of the partition function it was assumed that identical particles are distinguishable. However, gas particles are not distinguishable. It is sometimes convenient to choose $N$ equal to Avagadro's number and use a molar basis for the resulting calculations.

The Helmholtz free energy is given in terms of the partition function as

$$A = -kT \ln \frac{Z^N}{N!}. \tag{3-11}$$

The pressure and entropy are related to the partition function by

$$p = -\left.\frac{\partial A}{\partial V}\right)_T, \tag{3-12}$$

$$s = -\left.\frac{\partial A}{\partial T}\right)_V, \tag{3-13}$$

The internal energy may be written as

$$U = A + Ts \tag{3-14}$$

and the enthalpy and the Gibbs free energy as

$$H = U + pV, \tag{3-15}$$

$$F = A + pV. \tag{3-16}$$

Other thermodynamic functions can be obtained from combinations of the above equations. The specific heats can, of course, be obtained from the internal energy and enthalpy.

For a large collection of particles, as would be generally considered, Stirling's approximation for the factorial may be used:

$$\ln N! = N \ln N - N$$

and Eqs. (3–11)–(3–14) may be written, respectively, in terms of the partition function as

$$A = -NkT\left[\ln\frac{Z}{N} + 1\right], \tag{3-17}$$

$$p = NkT\frac{1}{Z}\left.\frac{\partial Z}{\partial V}\right)_T, \tag{3-18}$$

$$s = Nk\left[\ln\frac{Z}{N} + 1\right] + NkT\frac{1}{Z}\left.\frac{\partial Z}{\partial T}\right)_V, \tag{3-19}$$

$$U = NkT^2\frac{1}{Z}\left.\frac{\partial Z}{\partial T}\right)_V. \tag{3-20}$$

It can be seen that the evaluation of the thermodynamic functions reduces to the problem of the evaluation of the partition function. The evaluation of the translational, rotational, vibrational, and electronic partition functions will be discussed in the following sections. Explicit expressions for the partition functions of the atoms and molecules which are found in air are given in ref. 1.

## 3.4 Partition Functions for a Perfect Gas

### 3.4.1 Translational Partition Function

The Schrödinger wave equation can be solved for particles with only translational energy confined to a rectangular box. The total translational energy is found to be

$$E^{\mathrm{tr}} = \frac{h^2}{8m}\left(\frac{q_x^2}{a^2} + \frac{q_y^2}{b^2} + \frac{q_z^2}{c^2}\right) \tag{3-21}$$

where $x$, $y$, and $z$ are the space coordinates, $q_x$, $q_y$, and $q_z$ are the respective integer quantum numbers, $a$, $b$, and $c$ are the respective lengths of the rectangular box, $m$ is the mass of the particle, and $h$ is Planck's constant. Substituting Eq. (3–21) into Eq. (3–6), we obtain for the partition function

$$Z^{\mathrm{tr}} = \sum_{q_x q_y q_z} \exp\left[-\frac{h^2}{8mkT}\left(\frac{q_x^2}{a^2} + \frac{q_y^2}{b^2} + \frac{q_z^2}{c^2}\right)\right]. \tag{3–22}$$

The sum may be split into the product of three separate summations over values of $q_x$, $q_y$, and $q_z$, and each of these sums may be approximated by an integral and evaluated. The result is

$$\begin{aligned} Z^{\mathrm{tr}} &= Z_x Z_y Z_z \\ &= abc\left(\frac{2\pi mkT}{h^2}\right)^{3/2}. \end{aligned} \tag{3–23}$$

Since this result is not restricted to rectangular boxes, $abc$ can be replaced by the volume $V$ and the final result is

$$Z^{\mathrm{tr}} = V\left(\frac{2\pi mkT}{h^2}\right)^{3/2}. \tag{3–24}$$

### 3.4.2 Rotational Partition Function

Solving the Schrödinger wave equation for the case of a rigid linear rotator having a moment of inertia $I$, the allowable energy states are found to be given by

$$E^{\mathrm{rot}} = J(J+1)\frac{h^2}{8\pi^2 I} \tag{3–25}$$

where $J$ is the rotational quantum number which can have values of 0, 1, 2, 3, etc. The corresponding degeneracies are given by

$$g_J = 2J + 1. \tag{3–26}$$

Substituting Eqs. (3–25) and (3–26) into Eq. (3–4), the resulting partition function is

$$Z^{\text{rot}} = \sum_{j=0}^{\infty} (2J + 1) \exp\left[-J(J + 1) \frac{h^2}{8\pi^2 IkT}\right]. \tag{3-27}$$

It is convenient to let

$$\theta_{\text{rot}} = \frac{h^2}{8\pi^2 Ik} \tag{3-28}$$

where $\theta_{\text{rot}}$ is called the characteristic temperature for rotation, since it has the dimensions of a temperature. Then Eq. (3–27) becomes

$$Z^{\text{rot}} = \sum_{j=0}^{\infty} (2J + 1) \exp\left[-J(J + 1) \frac{\theta_{\text{rot}}}{T}\right] \tag{3-29}$$

which can be expressed as

$$Z^{\text{rot}} = 1 + 3 \exp\left(-\frac{2\theta_{\text{rot}}}{T}\right) + 5 \exp\left(-\frac{6\theta_{\text{rot}}}{T}\right) + 7 \exp\left(-\frac{12\theta_{\text{rot}}}{T}\right) + \cdots. \tag{3-30}$$

Equation (3–30) cannot be summed explicitly. However, for temperatures considerably higher than $\theta_{\text{rot}}$, the sum in Eq. (3–27) can be replaced by an integral and evaluated. The result is

$$Z^{\text{rot}} = \frac{8\pi^2 IkT}{\sigma h^2} \tag{3-31}$$

where $\sigma$ is the symmetry number. The symmetry number is equal to unity for a diatomic molecule consisting of two different atoms such as NO, and is equal to 2 for a diatomic molecule made up of identical atoms such as $N_2$.

The characteristic temperature for rotation is quite low for most gases (2.1 °K for oxygen), and therefore when dealing with ordinary temperatures and above, the rotational energy is fully excited and Eq. (3–31) can be used for the partition function. However, at low temperatures, Eq. (3–29) must be used. For more complicated molecules the evaluation of the rotational partition function is more difficult.

### 3.4.3 Vibrational Partition Function

Solving the Schrödinger wave equation for the case of a linear harmonic oscillator, the possible vibrational energies are found to be

$$E^{\mathrm{vib}} = (v + \tfrac{1}{2})h\nu \tag{3–32}$$

where $v$ is vibrational quantum number which can have values of 0, 1, 2, 3, etc., and $\nu$ is the vibrational frequency. The statistical weight of each vibrational energy level of different frequency is found to be unity. Therefore, the vibrational partition function is

$$Z^{\mathrm{vib}} = \sum_{v=0}^{\infty} \exp\left[-(v + \tfrac{1}{2})\frac{h\nu}{kT}\right]. \tag{3–33}$$

Defining the characteristic temperature for vibration as

$$\theta_{\mathrm{vib}} = h\nu/k \tag{3–34}$$

Eq. (3–33) becomes

$$Z^{\mathrm{vib}} = \sum_{v=0}^{\infty} \exp\left[-(v + \tfrac{1}{2})\frac{\theta_{\mathrm{vib}}}{T}\right]. \tag{3–35}$$

The sum of this series is

$$Z^{\mathrm{vib}} = \frac{\exp(-\theta_{\mathrm{vib}}/2T)}{1 - \exp(-\theta_{\mathrm{vib}}/T)}. \tag{3–36}$$

Some authors choose to evaluate the vibrational partition function differently since the above formulation does not result in the internal energy being zero at the absolute zero of temperature. Therefore, one subtracts from Eq. (3–32) the zero point energy, i.e., the energy of the lowest ($v = 0$) level which is equal to $\frac{1}{2}h\nu$. The result is

$$E^{\text{vib}} = (v + \tfrac{1}{2})h\nu - \tfrac{1}{2}h\nu \tag{3-37}$$
$$= vh\nu.$$

The vibrational partition function then becomes

$$\begin{aligned} Z^{\text{vib}} &= \sum_{v=0}^{\infty} \exp\left[-\frac{vh\nu}{kT}\right] \\ &= \sum_{v=0}^{\infty} \exp\left[-v\frac{\theta_{\text{vib}}}{T}\right]. \end{aligned} \tag{3-38}$$

The sum of this series is

$$Z^{\text{vib}} = \frac{1}{1 - \exp\left(-\theta_{\text{vib}}/T\right)}. \tag{3-39}$$

There is apparently no better reason for using Eq. (3–39) instead of Eq. (3–36) for the partition function except the one already given. A discussion of this point is given by Slater in ref. 2.

Like the rotational partition function, the evaluation of the vibrational partition function for polyatomic molecules is considerably more complicated.

### 3.4.4 Electronic Partition Function

When one of the electrons of an atom or molecule jumps from one state to a higher energy state, energy is absorbed, and this energy is subsequently released when the electron returns to its original state. At high temperatures, the population of the higher electronic states is significant. High temperatures in this case mean that $kT$ is of the order of the energy of the first excited state, the ground state, by definition, having zero energy.

Since the energy of the ground state is zero, the electronic partition function can be written as

$$\begin{aligned} Z^{\text{elec}} &= \sum_{m=0}^{\infty} g_m \exp\left(-\frac{E_m}{kT}\right) \\ &= g_0 + g_1 \exp\left(-\frac{E_1}{kT}\right) + g_2 \exp\left(-\frac{E_2}{kT}\right) + \cdots. \end{aligned} \tag{3-40}$$

The series of Eq. (3–40) is actually divergent because the energy differences between the ground state and the excited states approach a constant value, namely, the ionization energy, and there are an infinite number of bound states below this limit. Also, the statistical weight approaches infinity as the energy approaches the ionization energy. Therefore, the later terms of the infinite series do not diminish.

There are several ways to resolve this difficulty. Some investigators noted that for the higher energy states, the electrons have very large orbits which may approach the distance of separation of the gas molecules. Therefore, they cut off the series at the energy level corresponding to the orbit radius equal to half the mean distance between the molecules of the gas (see for example refs. 3 and 4). In ref. 5 it is pointed out that this is a difficult criterion to employ, since it depends explicitly on the ion density of the gas, which in turn will depend on the value of $Z^{\text{elec}}$ Thus, in practice, some sort of iterative procedure is required to evaluate $Z^{\text{elec}}$ by this method.

An alternate method of approximation which is used in ref. 5 is to cut off the series at values of the binding energy of the order of $kT$. Some authors cut off the series at some arbitrary number of terms for simplicity. This method is used, for example, in ref. 6. The error introduced by this method increases with increasing temperature. In any event, it can be seen that the sum of the series of Eq. (3–40), although infinite in a mathematical sense, can be evaluated in practice.

In cases where calculations of high accuracy are not required, it is possible to approximate the series of Eq. (3–40) with the first two terms. Although, as stated previously, the error introduced in this way increases with increasing temperature, it will be small provided that the energy of the second excited state $E_2$ is much greater than $kT$. This condition will be met for many atoms and molecules for most temperatures encountered in engineering problems. For this case, the partition function becomes

$$Z^{\text{elec}} = g_0 + g_1 \exp(-E_2/kT). \tag{3-41}$$

To illustrate how the electronic partition function is evaluated, we will consider the argon atom and ion. The first two energy levels and the ionizational energy of the argon atom are given in the tabulation (ref. 3).

| $g_m$ | $E_m/k$ (°K) |
|---|---|
| 1 | 0 |
| 60 | 162,500 |
| Ionization energy | 183,000 |

The electronic partition function is, therefore

$$Z_{\text{A}}^{\text{elec}} = 1 + 60 \exp\left(-\frac{162{,}500}{T}\right) + \cdots. \tag{3-42}$$

For values of $T \ll 162{,}500$ °K, the second term is very small, and the partition function is given to a reasonable degree of accuracy by

$$Z_{\text{A}}^{\text{elec}} = g_0 = 1. \tag{3-43}$$

The first four energy levels and the ionization energy of the singly ionized argon atom are given in the tabulation (ref. 3).

| $g_m$ | $E_m/k$ (°K) |
|---|---|
| 4 | 0 |
| 2 | 2062 |
| 2 | 156,560 |
| 8 | 190,550 |
| Ionization energy | 320,800 |

The electronic partition function is

$$Z_{\mathrm{A}^+}^{\mathrm{elec}} = 4 + 2\exp\left(-\frac{2062}{T}\right) + 2\exp\left(-\frac{156{,}560}{T}\right) + 8\exp\left(-\frac{190{,}550}{T}\right) + \cdots. \tag{3–44}$$

For $T \ll 156{,}560$ °K, all of the terms beyond the second are very small. However, the second term is appreciable except for very low temperatures. Therefore, Eq. (3–44) can be approximated by

$$Z_{\mathrm{A}^+}^{\mathrm{elec}} = 4 + 2\exp\left(-\frac{2062}{T}\right). \tag{3–45}$$

Reference 4 contains a table of two-term approximations of the electronic partition functions for some of the common elements and their ions. Reference 7 lists the energies of electronic states of all common atoms and ions.

## 3.5 Thermal Equation of State

The pressure exerted by a collection of particles is given by Eq. (3–18) in terms of the partition function. Evaluating this expression in terms of the partition functions given in Section 3.4 leads to (only $Z^{\mathrm{tr}}$ contributes to the pressure):

$$p_i = \frac{kT}{V} N_i. \tag{3–46}$$

This partial pressure represents the pressure exerted by the $i$th species. Using Dalton's law of partial pressures, the total pressure of a mixture of different species can be expressed as

$$p = \frac{kT}{V} \sum N_i. \tag{3–47}$$

For a gas undergoing a chemical reaction, this expression is somewhat cumbersome to use since $\Sigma N_i$ is a variable and $V$ is the total volume. Rewriting Eq. (3–47) in terms of mass density and the universal gas constant,

$$p = \rho R_0 T \frac{\Sigma N_i}{\Sigma N_i M_i}. \tag{3–48}$$

This equation represents the thermal equation of state for a mixture of $i$ species, with or without chemical reactions.

Equation (3–48) is readily specialized. For the case of the dissociation of a pure diatomic gas, it becomes

$$p = \rho \frac{R_0}{M_{A_2}} T(1 + \alpha) \tag{3–49}$$

where $\alpha$ is defined as the mass fraction of the dissociated species.

$$\alpha = \frac{N_A M_A}{N_{A_2} M_{A_2} + N_A M_A}. \tag{3–50}$$

For the case of the ionization (single) of a pure monatomic gas, it becomes

$$p = \rho \frac{R_0}{M_A} T(1 + \phi_1) \tag{3–51}$$

where $\phi$ is defined as the mass fraction of the ionic species:

$$\phi_1 = \frac{N_{A^+} M_A}{N_A M_A + N_{A^+} M_A}. \tag{3–52}$$

For the case of double ionization of a pure monatomic gas, Eq. (3–48) may be written as

$$p = \rho \frac{R_0}{M_A} T(1 + \phi_1 + 2\phi_2) \tag{3–53}$$

where the $\phi$'s are defined as the mass fractions of the ionic species:

$$\phi_1 = \frac{N_{A^+} M_A}{N_A M_A + N_{A^+} M_A + N_{A^{++}} M_A}, \tag{3–54}$$

$$\phi_2 = \frac{N_{A^{++}} M_A}{N_A M_A + N_{A^+} M_A + N_{A^{++}} M_A}. \tag{3–55}$$

[In Eqs. (3–51) to (3–55) it was assumed that the molecular weight of the ions and atoms were identical.]

## 3.6 Caloric Equation of State

The caloric equation of state may be an equation for the internal energy, the enthalpy, or the specific heat. The enthalpy is readily calculated by combining Eqs. (3–15), (3–18), and (3–20). Thus, the enthalpy of the $i$th species is given by

$$H_i = N_i kT \left[ T \left.\frac{\partial \ln Z}{\partial T}\right)_V + V \left.\frac{\partial \ln Z}{\partial V}\right)_T \right]. \tag{3–56}$$

Expanding Eq. (3–56) using the partition functions given in Section 3.4 yields

$$H_i = \tfrac{5}{2} N_i kT + N_i kT^2 \frac{d}{dT} [\ln Z_{\text{rot}} + \ln Z_{\text{vib}} + \ln Z_{\text{elec}}]. \tag{3–57}$$

Inspection of this equation shows that the enthalpy of a classical smooth sphere kinetic theory model gas (no rotational, vibrational, or electronic contributions) is

$$H_i = \tfrac{5}{2} N_i kT$$

corresponding to the classical result for a monatomic gas. Similarly, including a rotational contribution but excluding vibrational and electronic effects, the result is

$$H_i = \tfrac{7}{2} N_i kT.$$

Again this corresponds the classical case of a diatomic gas.

In general, Eq. (3–57) can be expanded further, resulting in the following expression for the enthalpy of the $i$th species:

$$H_i = \tfrac{7}{2} N_i kT + N_i k\theta_v \frac{\exp(-\theta_v/T)}{1 - \exp(-\theta_v/T)} + N_i kT^2 \frac{d}{dT} \ln \sum g_m \exp\left(-\frac{E_m}{kT}\right). \tag{3–58}$$

Of course, if a monatomic gas is being considered as opposed to a diatomic gas, the rotational energy, $N_i kT$, and the vibrational term must be subtracted from Eq. (3–58).

For a dissociating gas, a mixture of atoms and molecules, such as that for which the thermal equation of state is given by Eq. (3–49), the enthalpy per unit mass is

$$h = \frac{(7-3\alpha)}{4}\frac{R_0}{M_A}T + \frac{(1-\alpha)}{2}\frac{R_0}{M_A}\theta_v\frac{\exp(-\theta_v/T)}{1-\exp(-\theta_v/T)} \tag{3–59}$$
$$+ \alpha D + \frac{R_0}{M_A}T^2\left[\alpha\frac{d}{dT}\ln\sum g_{m_A}\exp\left(-\frac{E_{m_A}}{kT}\right)\right.$$
$$\left. + \frac{(1-\alpha)}{2}\frac{d}{dT}\ln\sum g_{m_{A_2}}\exp\left(-\frac{E_{m_{A_2}}}{kT}\right)\right]$$

where $D$ is the dissociation energy.

Similarly, for a singly ionized monatomic gas, the specific enthalpy is (noting that the internal energy of an ionized monatomic gas is the sum of the translational energy, ionization energy, and electronic energy)

$$h = \frac{5(1+\phi)}{2}\frac{R_0}{M_A}T + \phi I \tag{3–60}$$
$$+ \frac{R_0}{M_A}T^2\left[(1-\phi)\frac{d}{dT}\ln\sum g_{m_A}\exp\left(-\frac{E_{m_A}}{kT}\right)\right.$$
$$\left. + \phi\frac{d}{dT}\ln\sum g_{m_{A^+}}\exp\left(-\frac{E_{m_{A^+}}}{kT}\right)\right]$$

with the same assumptions as were required for Eq. (3–51).

This specialization may be readily extended to cases of multiple ionization or combined dissociation and ionization mixtures. The primary difficulty is that the simplified analysis considered here will not be valid for high electron densities since the interparticle interactions have been neglected. For moderate or low ionization fractions, some simplifications for the above equations are possible and these approximations are discussed in Section 3.8.

## 3.7 Thermodynamic Equilibrium Condition

For a gas which does not have constant composition, i.e., a gas in which chemical reactions occur, a relation is required which expresses the amounts of the reactants present under equilibrium conditions as a function of temperature and pressure. This relation can be derived from statistical thermodynamics in terms of the partition functions of the reactants. For the reaction

$$\mathrm{AB} \leftrightarrows \mathrm{A} + \mathrm{B} \tag{3-61}$$

which can represent a dissociation or an ionization reaction, by using Eq. (3-17) and the condition for thermodynamic equilibrium, it can be shown that (see for example ref. 8)

$$\frac{N_{\mathrm{A}} N_{\mathrm{B}}}{N_{\mathrm{AB}}} = \frac{Z_{\mathrm{A}} Z_{\mathrm{B}}}{Z_{\mathrm{AB}}} \exp\left(-\frac{E_0}{kT}\right) \tag{3-62}$$

where the $Z$'s are the total partition functions and $E_0$ is the total energy change of the reaction. For example, if Eq. (3-61) represents a dissociation reaction, $E_0$ is the dissociation energy. The left-hand side of Eq. (3-62) is often referred to as the equilibrium constant.

If Eq. (3-10) is substituted into Eq. (3-62), the result is

$$\frac{N_{\mathrm{A}} N_{\mathrm{B}}}{N_{\mathrm{AB}}} = \frac{Z_{\mathrm{A}}^{\mathrm{tr}} Z_{\mathrm{B}}^{\mathrm{tr}}}{Z_{\mathrm{AB}}^{\mathrm{tr}}} \frac{Z_{\mathrm{A}}^{\mathrm{int}} Z_{\mathrm{B}}^{\mathrm{int}}}{Z_{\mathrm{AB}}^{\mathrm{int}}} \exp\left(-\frac{E_0}{kT}\right). \tag{3-63}$$

Substituting Eq. (3-24) for the translational partition function in Eq. (3-63) and simplifying leads to

$$\frac{N_{\mathrm{A}} N_{\mathrm{B}}}{N_{\mathrm{AB}}} = V \left(\frac{2\pi kT}{h^2}\right)^{3/2} \left(\frac{m_{\mathrm{A}} m_{\mathrm{B}}}{m_{\mathrm{AB}}}\right)^{3/2} \frac{Z_{\mathrm{A}}^{\mathrm{int}} Z_{\mathrm{B}}^{\mathrm{int}}}{Z_{\mathrm{AB}}^{\mathrm{int}}} \exp\left(-\frac{E_0}{kT}\right). \tag{3-64}$$

Since

$$\frac{N_{\mathrm{A}} N_{\mathrm{B}}}{N_{\mathrm{AB}} V} = \frac{n_A n_{\mathrm{B}}}{n_{\mathrm{AB}}} \tag{3-65}$$

where $n$ is the number of particles per unit volume, Eq. (3–64) becomes

$$\frac{n_{\mathrm{A}}\, n_{\mathrm{B}}}{n_{\mathrm{AB}}} = \left(\frac{2\pi kT}{h^2}\right)^{3/2} \left(\frac{m_{\mathrm{A}}\, m_{\mathrm{B}}}{m_{\mathrm{AB}}}\right)^{3/2} \frac{Z_{\mathrm{A}}^{\mathrm{int}} Z_{\mathrm{B}}^{\mathrm{int}}}{Z_{\mathrm{AB}}^{\mathrm{int}}} \exp\left(-\frac{E_0}{kT}\right). \tag{3–66}$$

For the set of ionization reactions for a monatomic gas

$$A_i \rightleftarrows A_{i+1} + e - I_i \qquad i = 0, 1, 2, \ldots$$

where $A_i$ is the particle ionized, $A_{i+1}$ is the same particle minus an electron, and $I_i$ is the ionization energy of the reaction, Eq. (3–66) becomes (making the usual approximation that $m_{i+1} = m_i$)

$$K_{i+1} = \frac{n_{i+1}\, n_e}{n_i} = \left(\frac{2\pi m_e\, kT}{h^2}\right)^{3/2} \frac{2Z_{i+1}^{\mathrm{elec}}}{Z_i^{\mathrm{elec}}} \exp\left(-\frac{I_i}{kT}\right) \tag{3–67}$$

where $K_{i+1}$ is the equilibrium constant for the reaction. There is one such equation for each reaction. It is noted that the electronic partition is used in Eq. (3–67), since for a monatomic gas, $Z^{\mathrm{int}}$ reduces to $Z^{\mathrm{elec}}$. The electronic partition function of a free electron reduces to its statistical weight, which, because of its two possible orientations of spin, is equal to 2.

For the ionization of a neutral atom, i.e.,

$$A_0 \rightleftarrows A_1 + e^- \tag{3–68}$$

Eq. (3–67) becomes

$$K_1 = \frac{n_1\, n_e}{n_0} = \frac{\phi_1^{\,2}}{1 - \phi_1} n_H = \left(\frac{2\pi m_e\, kT}{h^2}\right)^{3/2} \frac{2Z_1^{\mathrm{elec}}}{Z_0^{\mathrm{elec}}} \exp\left(-\frac{I_1}{kT}\right). \tag{3–69}$$

Now, the perfect gas pressure for this case is given by

$$p = (1 + \phi_1) n_H\, kT. \tag{3–70}$$

Therefore, Eq. (3–69) becomes

$$\frac{\phi_1^{\,2}}{1 - \phi_1^{\,2}} = \left(\frac{2\pi m_e}{h^2}\right)^{3/2} \frac{(kT)^{5/2}}{p} \frac{2Z_1^{\mathrm{elec}}}{Z_0^{\mathrm{elec}}} \exp\left(-\frac{I_1}{kT}\right). \tag{3–71}$$

Equation (3–69) or (3–71) is called the Saha equation. Equation (3–67) might be called the generalized Saha equation. Corresponding equations can easily be derived for each successive ionization reaction, that is, for the ionization of a singly ionized atom (double ionization), for the ionization of a doubly ionized atom (triple ionization), etc.

For the ionization reaction of Eq. (3–68), an equilibrium equation can be derived from classical thermodynamics, namely,

$$\frac{\phi^2}{1-\phi^2} = C\frac{T^{5/2}}{p}\exp\left(-\frac{I}{kT}\right). \tag{3–72}$$

It can be seen that Eq. (3–72) is similar to Eq. (3–71). The constant $C$ cannot be evaluated from classical thermodynamics. $C$ is sometimes taken as the product of the constant terms of Eq. (3–71), so Eq. (3–72) becomes

$$\frac{\phi^2}{1-\phi^2} = \left(\frac{2\pi m_e}{h^2}\right)\frac{(kT)^{5/2}}{p}\exp\left(-\frac{I}{kT}\right). \tag{3–73}$$

Equation (3–73) is known as the Saha equation, since it was Saha (ref. 9) who first obtained this expression, although the same name is also given to Eq. (3–71). The internal partition functions, of course, do not appear in the derivation from classical thermodynamics. It is probably for this reason, and also because Saha himself used it, that Eq. (3–73) is sometimes given as the correct equilibrium relation for the ionization reaction. It is only correct, however, if

$$\frac{2Z_{i+1}^{\text{elec}}}{Z_i^{\text{elec}}} = 1$$

which, while possible, is not generally true.

For argon, for example, using Eqs. (3–43) and (3–45), we have

$$\frac{2Z_1^{\text{elec}}}{Z_0^{\text{elec}}} = \frac{(2)\,[4+2\exp(-2062/T)]}{1}. \tag{3–74}$$

Even at very low temperatures where electronic excitation is not important, the smallest value Eq. (3–74) can have is 8. It can be seen, therefore, that the use of Eq. (3–73) can lead to a significant error.

## 3.8 Approximate Equations

Lighthill (ref. 10) has suggested a set of approximate equations describing an "ideal dissociating gas" that have been widely used. Lighthill's approximations are limited to the case of the dissociation of a diatomic molecule. He assumed that the enthalpy of the molecular species could be adequately represented by neglecting the electronic contributions entirely and setting the vibrational contribution equal to $\frac{1}{2}N_i kT$. Similarly, the atomic species can be considered to possess no electronic energy. With this analysis Eq. (3–59) may be written as

$$h = \frac{(9+\alpha)}{4}\frac{R_0}{M_{\mathrm{A}}}T + \alpha D. \tag{3–75}$$

In addition, the condition for equilibrium, Eq. (3–66), can be approximated as

$$\frac{\alpha^2}{1-\alpha} = \frac{\rho_{\mathrm{D}}}{\rho}\exp\left(-\frac{E_0}{kT}\right). \tag{3–76}$$

The characteristic density, $\rho_{\mathrm{D}}$, defined as

$$\rho_{\mathrm{D}} = \frac{M_{\mathrm{A}}}{2N_0}\left(\frac{2\pi kT}{h^2}\right)^{3/2} m_{\mathrm{A}}^{3/2}\frac{Z_{\mathrm{A}}^{\mathrm{int}\,2}}{Z_{\mathrm{A}_2}^{\mathrm{int}}} \tag{3–77}$$

is assumed to be a constant since it varies little with temperature in the range of temperatures generally considered for the dissociation of atmospheric gases. A slightly less restrictive approach has been used by Anderson (ref. 11) with very good results but reaffirming Lighthill's approximation, since, although the vibrational term was allowed to vary with temperature in both the enthalpy and the equilibrium equations, no significant effect was noted on the flow properties.

A similar approximation for a singly ionized gas has been suggested by Duclos, Aeschliman, and Cambel (ref. 12). They studied the simplification of the Saha equation and the expression for the internal energy and found suitable arrangements of the partition functions which would remain substantially constant over relatively wide ranges of temperature. Equation (3–69) can be written as follows:

$$K_1 = \frac{n_1 n_e}{n_0} = \left(\frac{2\pi m_e k}{h^2}\right)^{3/2} c T^{(3/2)-q} \exp\left(-\frac{I_1}{kT}\right) \tag{3-78}$$

where

$$c = \frac{2Z_1^{\text{elec}}}{Z_0^{\text{elec}}} T^q.$$

To exactly duplicate Lighthill's approximation, it must be shown that the entire right-hand side of Eq. (3–78), with the exception of the exponential term, is a constant. This is not possible for nitrogen and oxygen, but a value of the exponent $q$ can be found which will yield an essentially constant $c$.

For nitrogen, using the first few terms of the partition functions given by Martinek (ref. 13), a value of $q = \frac{1}{4}$ is found to give values of $c$, shown in Table 3–1, which lie within 13% of the average value of 35.35. Therefore, the following approximation can be suggested for singly ionized nitrogen in the temperature range 5000° to 30,000 °K:

$$K_1 = \frac{n_1 n_e}{n_0} = 35.35 \left(\frac{2\pi m_e k}{h^2}\right)^{3/2} T^{5/4} \exp\left(-\frac{I_1}{kT}\right). \tag{3-79}$$

For oxygen, using energy levels given by Moore (ref. 7), it is found that in order for $c$ to be insensitive to temperature, one should set $q = -\frac{1}{4}$. The values of $c$ at different temperatures are shown in Table 3–2. The average value is 0.0953, and the maximum deviation is 12.3% at 5000 °K. Hence, for singly ionized oxygen between 5000° and 30,000 °K, the approximate expression for the equilibrium constant may be written as follows:

$$K_1 = \frac{n_1 n_e}{n_0} = 0.0953 \left(\frac{2\pi m_e k}{h^2}\right)^{3/2} T^{7/4} \exp\left(-\frac{I_1}{kT}\right). \qquad (3\text{–}80)$$

To investigate possible simplifications of the expression for the internal energy of a monatomic gas, it is first written as follows:

$$U = kT\left\{\left[\frac{3}{2} + T\frac{\partial \ln Z_0^{\text{elec}}}{\partial T}\right)_V\right] N_0 + \left[3 + T\frac{\partial \ln Z_1^{\text{elec}}}{\partial T}\right)_V\right] N_1\right\} + I_1 N_1. \qquad (3\text{–}81)$$

For a convenient approximation the quantities within the brackets should behave as constants. However, Martinek's calculations show that this is not the case for nitrogen. On the other hand, Martinek has shown that the term $T\,\partial \ln Z^{\text{elec}}/\partial T)_V$ never exceeds 0.7 for the atomic species and 0.3 for the ions. In other words, the electronic energy is small in the 4000° to 40,000 °K range in comparison with the translational energy. Therefore, to a fair degree of approximation, the quantities in the brackets of Eq. (3–81) can be replaced by their mean values. Consequently, Eq. (3–81) may be written as follows for singly ionized atomic nitrogen:

$$U = kT(1.85\,N_0 + 3.15\,N_1) + I_1 N_1. \qquad (3\text{–}82)$$

Calculations of the values of $T\,\partial \ln Z^{\text{elec}}/\partial T)_V$ indicate that a similar approximation can also be made for oxygen. For temperatures between 5000° and 20,000 °K, it is found that the internal energy of singly ionized atomic oxygen can be given approximately by

$$U = kT(1.66\,N_0 + 3.36\,N_1) + I_1 N_1. \qquad (3\text{–}83)$$

If an expression for the internal energy of nitrogen and oxygen molecules is desired, the appropriate expressions for the rotational and vibrational energies must be added to Eqs. (3–82) and (3–83). It is quite possible that similar approximations can be made for other gases, and for higher degrees of ionization.

TABLE 3–1 NITROGEN

| $T$(°K) | $c = 2T^{1/4}Z_1^{\text{elec}}/Z_0^{\text{elec}}$ |
|---|---|
| 5,000 | 33.80 |
| 10,000 | 40.02 |
| 15,000 | 37.70 |
| 20,000 | 34.70 |
| 25,000 | 33.10 |
| 30,000 | 32.80 |

TABLE 3–2 OXYGEN

| $T$(°K) | $c = 2T^{1/4}Z_1^{\text{elec}}/Z_0^{\text{elec}}$ |
|---|---|
| 5,000 | 0.107 |
| 10,000 | 0.090 |
| 15,000 | 0.090 |
| 20,000 | 0.091 |
| 25,000 | 0.096 |
| 30,000 | 0.098 |

## 3.9 Double Ionization

In the past, most investigators have assumed that, at atmospheric pressure and at temperatures below about 25,000 °K, the number of doubly ionized atoms is negligible. However, recent calculations of ionization behind shock waves in argon by Venable (ref. 14) have shown that, at about 25,000 °K and atmospheric pressure, $\phi_2/\phi_1$ is over $\frac{1}{2}$ while at a pressure of about $10^{-3}$ mm Hg, $\phi_2/\phi_1$ is about 2 for temperatures as low as 10,000 °K. Clearly, double ionization cannot be neglected arbitrarily.

For the case when both single and double ionization occurs, their respective equilibrium constants $K_1$ and $K_2$ are as follows:

$$K_1 = \frac{n_1 n_e}{n_0} = \frac{\phi_1(\phi_1 + 2\phi_2)}{(1 - \phi_1 - \phi_2)} n_H = \left(\frac{2\pi m_e kT}{h^2}\right)^{3/2} \frac{2Z_1^{\text{elec}}}{Z_0^{\text{elec}}} \exp\left(-\frac{I_1}{kT}\right), \tag{3–84}$$

$$K_2 = \frac{n_2 n_e}{n_1} = \frac{\phi_2}{\phi_1}(\phi_1 + 2\phi_2)\, n_H = \left(\frac{2\pi m_e kT}{h^2}\right)^{3/2} \frac{2Z_2^{\text{elec}}}{Z_1^{\text{elec}}} \exp\left(-\frac{I_2}{kT}\right). \tag{3–85}$$

The amount of double ionization can, of course, be determined by solving Eqs. (3–84) and (3–85) for $\phi_1$ and $\phi_2$ for the pressure and temperature ranges of interest. However, it is possible to determine if double ionization should be considered in a particular problem by a simpler, although equivalent, method.

Consider the set of equations from which the numbers of singly and doubly ionized particles can be determined:

$$n_H = n_0 + n_1 + n_2, \qquad n_e = n_1 + 2n_2$$

$$\frac{n_1 n_e}{n_0} = K_1, \qquad \frac{n_2 n_e}{n_1} = K_2.$$

It is desired to find the condition for which the number of doubly ionized particles equals some fraction $f$ of the number of singly ionized particles, i.e.,

$$n_2 = f n_1 \tag{3–86}$$

where $f$ would be much less than unity. For $f \leqslant 0.1$, the error involved in neglecting double ionization should be small. Using Eq. (3–86) and eliminating $n_e$, $n_0$, and $n_1$ from the above equations leads to

$$K_2 = \frac{f(1+f)K_1}{2}\left[-1 + \left(1 + 4\frac{(1+2f)}{(1+f)^2}\frac{n_H}{K_1}\right)^{1/2}\right]. \tag{3–87}$$

If $K_1$, $K_2$, and $n_H$ are calculated for the condition for which double ionization is expected to be the greatest, and $K_1$ and $n_H$ are substituted in Eq. (3–87), the corresponding value of $K_2$ will be determined for which the chosen condition, Eq. (3–86), will hold. If the actual $K_2$ is lower than the value calculated from Eq. (3–87), then double ionization can be neglected. Equations which determine the importance of triple, quadruple, etc., ionization can, of course, be derived in a similar manner.

## 3.10 List of Symbols

Latin symbols:

$A$ — Helmholtz free energy
A — atomic species
AB — diatomic species
B — atomic species
$c$ — simplifying constant in equation for $K$
$C$ — constant
$D$ — dissociation energy
$e$ — electron
$E$ — energy of particle
$F$ — Gibbs free energy
$g$ — statistical weight
$h$ — Planck's constant; also, specific enthalpy
$H$ — enthalpy
$I$ — ionization energy; also, molecular moment of inertia
$J$ — rotational quantum number
$k$ — Boltzmann's constant
$K$ — equilibrium constant
$m$ — particle mass
$M$ — molecular weight
$n$ — number density
$N$ — number of particles
$N_0$ — Avagadro's number
$p$ — pressure
$R_0$ — universal gas constant
$s$ — entropy
$T$ — temperature
$U$ — internal energy
$v$ — vibrational quantum number
$V$ — volume
$x, y, z$ — space coordinates
$Z$ — partition function

Greek symbols:

$\alpha$ — mass fraction of atomic species
$\theta$ — characteristic temperature
$\nu$ — characteristic frequency of vibration
$\rho$ — mass density
$\rho_D$ — characteristic density
$\sigma$ — symmetry number
$\phi$ — mass fraction of ionic species

Subscripts:

A — atomic species
$A_2$ — diatomic species
$A^+$ — singly ionized atom
$A^{++}$ — doubly ionized atom
$e$ — electron
$H$ — heavy particles (atoms and ions)
$i, j, k$ — summation indices
$m$ — summation index
rot — rotational
vib — vibrational
0 — neutral particle
1 — singly ionized particle
2 — doubly ionized particle

Superscripts:

elec — electronic
int — internal, including rotational, vibrational, and electronic excitation
rot — rotational
tr — translational
vib — vibrational
$q$ — exponent of temperature in equation for $c$

## References

1. Logan, J. G., Jr., "The Calculation of the Thermodynamic Properties of Air at High Temperatures." Cornell Aeronautical Lab., Report AD–1052–A–1, May, 1956.
2. Slater, J. C., *Introduction to Chemical Physics*. McGraw-Hill, New York, 1939.
3. Bond, J. W., Jr., "The Structure of a Shock Front in Argon." Los Alamos Scientific Lab., Report LA–1693, July 1, 1954.
4. Laporte, O., "High Temperature Shock Waves," in *Combustion and Propulsion, Third Agard Colloquium*. Pergamon Press, New York, 1958, pp. 499–524.
5. Benson, S. W., J. H., Buss, and H., Myers, "Thermodynamic Properties of Ionized Gases." IAS Paper No. 59–95, Presented at the IAS National Summer Meeting, Los Angeles, June 16–19, 1959.
6. Gilmore, F. R., "Equilibrium Composition and Thermodynamic Properties of Air to 24,000 °K." The Rand Corp., Report RM–1543, August 24, 1955.
7. Moore, C. E., "Atomic Energy Levels." N.B.S. circular 467: Vol. I, Elements 1–23, June 15, 1949; Vol. II, Elements 24–41, August 15, 1952; Vol. III, Elements 42–57 and 72–89, May 1, 1958.
8. Duclos, D. P., "The Equation of State of an Ionized Gas." (Gas Dynamics Lab. Report). AEDC TN 60–192, October, 1960.
9. Saha, M. N., "Ionization in the Solar Chromosphere." *Philosophical Magazine*, Vol. 40, No. 238, October, 1920, pp. 472–488.
10. Lighthill, M. J., "Dynamics of a Dissociating Gas." ARC 18,837, November 14, 1956.
11. Anderson, T. P., "The Effect of Recombination Rate on the Flow of a Dissociating Diatomic Gas" (Gas Dynamics Lab., Report). AEDC TR 61–12, September, 1961.
12. Duclos, D. P., D. P. Aeschliman, and A. B. Cambel, "Approximate Equation for Perfect Gas Plasma." *ARS Journal*, Vol. 32, No. 4, April, 1962, pp. 641–642.
13. Martinek, F., "Thermodynamic and Electrical Properties of Nitrogen at High Temperatures," in *Thermodynamic and Transport Properties of Gases, Liquids and Solids* (edited by Y. S. Touloukian). McGraw-Hill, New York, 1959, pp. 130–156.
14. Venable, D., Unpublished work, Los Alamos Scientific Lab., 1959.

*Chapter 4*

# The Debye-Hückel Theory for Ionized Gases

## 4.1 *Introduction*

In 1923, Debye and Hückel (ref. 1) published their very important theory of electrolytes. Probably the first person who suggested the application of their theory to determine the properties of ionized gases was Rosseland (ref. 2) in 1924. The similarities between an electrolyte solution and an ionized gas are apparent. The former consists of a solute made up of positive and negative ions, as well as neutral molecules, immersed in a solvent of dielectric constant $D$, the most common solvent being water. On the other hand, an ionized gas consists of positive ions, electrons, and neutral particles with free space taking the place of the solvent. The value of the dielectric constant used throughout this discussion is the value for free space.

## 4.2 *Fundamental Equations*

Debye and Hückel started with Poisson's equation from which, in principle, the electric potential can be determined for any region in which charges are distributed. Poisson's equation is

$$\nabla^2 \phi = -\frac{4\pi}{D} \rho \tag{4-1}$$

where $\rho$ is the charge density. (The factor $4\pi$ appears since the cgs system of units is used.) The charge density is the number of electronic charges per unit volume and is given by

$$\rho = \sum_i n_i' \, z_i \, \varepsilon \quad \text{(4–2)}$$

where $\varepsilon$ is the electronic charge, and $z_i$ is the valence of the charged particle.

The potential at a distance $r$ from the $j$ ion can, therefore, be obtained from the solution of

$$\nabla^2 \phi_j(r) = -\frac{4\pi}{D}\rho(r) = -\frac{4\pi}{D}\sum_i n_i'(r) z_i \, \varepsilon. \quad \text{(4–3)}$$

It is assumed that the distribution of the charged particles about a given ion is given by Boltzmann's formula

$$n_i'(r) = n_i \exp\left(- W_{ji}/kT\right) \quad . \text{(4–4)}$$

where $n_i$ is the average density of particles of species $i$ and $W_{ji}$ is the potential energy of the $i$th ion in the vicinity of the central $j$th ion. The fundamental assumption of Debye and Hückel was to replace $W_{ji}$ by $z_i \, \varepsilon \phi_j$. This approximation employs the principle of linear superposition of fields, that is, the potential is assumed to be proportional to the charge. The validity of this assumption, while somewhat in doubt, appears to be valid for low charged particle densities. A detailed discussion of this point is given by Fowler and Guggenheim (ref. 3).

Poisson's equation therefore becomes

$$\nabla^2 \phi_j(r) = -\frac{4\pi\varepsilon}{D}\sum_i n_i \, z_i \exp\left(-\frac{z_i \, \varepsilon \phi_j}{kT}\right). \quad \text{(4–5)}$$

In order to solve this equation conveniently, Debye and Hückel expanded the exponential in a power series:

$$\exp\left(-\frac{z_i \, \varepsilon\phi_j}{kT}\right) = 1 - \frac{z_i \, \varepsilon\phi_j}{kT} + \frac{1}{2!}\left(\frac{z_i \, \varepsilon\phi_j}{kT}\right)^2 - \cdots. \quad \text{(4–6)}$$

They made the further approximation that $z_i \, \varepsilon\phi_j$ is small compared to $kT$ and therefore the series could be replaced by the first two terms. Robinson and Stokes (ref. 4) have observed that this approximation is quite accurate for an electrolyte in which the positive and negative ions have the same magnitude of charge. In this case, when Eq. (4–6) is substituted into Eq. (4–5), the even powers of $\phi_j$ vanish. Equation (4–5) thus becomes

$$\nabla^2 \phi_j(r) = -\frac{4\pi\varepsilon}{D} \sum_i n_i z_i + \frac{4\pi\varepsilon^2 \phi_j}{DkT} \sum_i n_i z_i^2. \tag{4–7}$$

Now, since the gas is electrically neutral, the sum in the first term on the right-hand side is zero. Letting

$$\varkappa^2 = \frac{4\pi\varepsilon^2}{DkT} \sum_i n_i z_i^2 \tag{4–8}$$

and using the spherically symmetric form of the Laplacian operator, Eq. (4–7) becomes

$$\nabla^2 \phi_j = \frac{1}{r^2} \frac{d}{dr} \left( r^2 \frac{d\phi_j}{dr} \right) = \varkappa^2 \phi_j \tag{4–9}$$

which is the basic equation of the Debye-Hückel theory. (The physical meaning of the quantity $\varkappa$ is discussed in Section 4.5.)

The general solution of Eq. (4–9) is

$$\phi_j(r) = C' \frac{e^{-\varkappa r}}{r} + C'' \frac{e^{\varkappa r}}{r}. \tag{4–10}$$

Since the potential must remain finite at $r = \infty$, it follows that $C'' = 0$. The constant $C'$ can be determined from the condition that the electric induction, $d\phi/dr$, be continuous at the boundary between the ion $j$ and the surrounding atmosphere (see, e.g., ref. 3). An alternate method is to note that, since the gas is electrically neutral, the total charge about the ion $j$ must equal $-z_j \varepsilon$, i.e.,

$$\int_a^\infty 4\pi r^2 \rho_j(r) dr = -z_j \varepsilon. \tag{4–11}$$

Substituting the charge density

$$\rho_j(r) = -\frac{D\varkappa^2}{4\pi}\left(C'\frac{e^{-\varkappa r}}{r}\right) \tag{4-12}$$

in Eq. (4–11), evaluating the integral, and solving for $C'$ leads to

$$C' = \frac{z_j \varepsilon}{D} \frac{e^{\varkappa a}}{1 + \varkappa a}. \tag{4-13}$$

The potential is then given by

$$\phi_j(r) = \frac{z_j \varepsilon}{D} \frac{e^{\varkappa a}}{1 + \varkappa a} \frac{e^{-\varkappa r}}{r}. \tag{4-14}$$

The parameter $a$ is the distance of closest approach of charged particles, that is, it is shortest possible distance between the center of ion $j$ and the center of ion $i$. If $a_j$ and $a_i$ are the effective diameters of the $j$ and $i$ ions, respectively, then

$$a_{ij} = \tfrac{1}{2}(a_i + a_j). \tag{4-15}$$

For electrolytes, it is usually assumed that $a_{ij}$ has the constant value $a$ for all ions. This point is discussed in relation to ionized gases later in this chapter.

Subtracting the self-potential of the $j$ ion, $z_j \varepsilon/Da$, that is, the potential at $r = a$ due to the $j$ ion itself, from the value of Eq. (4–14) at $r = a$, leads to

$$\phi_j{}^*(a) = \frac{z_j \varepsilon}{Da} \frac{1}{1 + \varkappa a} - \frac{z_j \varepsilon}{Da}$$

$$= -\frac{z_j \varepsilon}{D} \frac{\varkappa}{1 + \varkappa a}. \tag{4-16}$$

This result represents the potential at the ion $j$ due to all of the other charged particles in the gas. Therefore, Eq. (4–16) is used in the derivation of the electrostatic contributions to the thermodynamic properties. It should be noted that the Debye-Hückel potential represents only the effects of the long-range Coulomb forces. Short-range forces, which are important at high densities, are not included.

## 4.3 Thermodynamic Properties

According to the laws of electrostatics, the energy of the electrical interaction of a system of charged particles is

$$E = \tfrac{1}{2} \sum_i N_i z_i \varepsilon \phi_i. \tag{4–17}$$

Substituting Eq. (4–16) into Eq. (4–17) leads to

$$E^{\mathrm{DH}} = -\frac{\varepsilon^2 \varkappa \Sigma_i N_i z_i^2}{2D} \frac{1}{1 + \varkappa a}. \tag{4–18}$$

It is well known from thermodynamics that

$$\left.\frac{\partial (F/T)}{\partial T}\right)_{V, N_i} = -\frac{E}{T^2} \tag{4–19}$$

where $F$ is the Helmholtz free energy. Integrating this relation at constant volume gives

$$\frac{F^{\mathrm{DH}}}{T} = -\int \frac{E^{\mathrm{DH}}}{T^2} dT + \text{const.}$$

$$= \int \frac{\varepsilon^2 \varkappa \Sigma_i N_i z_i^2}{2D} \frac{1}{1 + \varkappa a} dT + \text{const.} \tag{4–20}$$

Using the condition that when $T \to \infty$, $F^{\mathrm{DH}}/T \to 0$, to evaluate the constant, the result is

$$F^{\mathrm{DH}} = -\frac{\varepsilon^2 \varkappa \Sigma_i N_i z_i^2}{3D} \frac{3}{(a\varkappa)^3} [\tfrac{1}{2}(a\varkappa)^2 - a\varkappa + \ln (a\varkappa + 1)]. \tag{4–21}$$

For $a\varkappa \leqslant 1$, $\ln (a\varkappa + 1)$ can be expanded in a power series, and Eq. (4–21) becomes

$$F^{\mathrm{DH}} = -\frac{\varepsilon^2 \varkappa \Sigma_i N_i z_i^2}{3D} \left[1 - \frac{3}{4}(a\varkappa) + \frac{3}{5}(a\varkappa)^2 - \frac{3}{6}(a\varkappa)^3 + \cdots\right]. \tag{4–22}$$

Using the thermodynamic relation

$$p = -\left.\frac{\partial F}{\partial V}\right)_{T, N_i} \tag{4-23}$$

the electrical contribution to the pressure is found to be

$$p^{\text{DH}} = -\frac{\varepsilon^2 \varkappa \sum_i n_i z_i^2}{6D} \frac{6}{(a\varkappa)^2}\left[\frac{1}{2} + \frac{1}{2}\frac{1}{a\varkappa + 1} - \frac{1}{a\varkappa}\ln(a\varkappa + 1)\right]. \tag{4-24}$$

For $a\varkappa < 1$, $(a\varkappa + 1)^{-1}$ and $\ln(a\varkappa + 1)$ can be expanded in power series and Eq. (4-24) becomes

$$p^{\text{DH}} = -\frac{\varepsilon^2 \varkappa \sum_i n_i z_i^2}{6D}\left[1 - \frac{3}{2}(a\varkappa) + \frac{9}{5}(a\varkappa)^2 - 2(a\varkappa)^3 + \dots\right]. \tag{4-25}$$

For very low concentrations of charged particles, it is a good approximation to let $a = 0$ in the preceding equations. The resulting expressions are called the Debye-Hückel limiting laws:

$$E^{\text{DH}} = -\frac{\varepsilon^2 \varkappa}{2D}\sum_i N_i z_i^2 = -\frac{kTV\varkappa^3}{8\pi}, \tag{4-26}$$

$$F^{\text{DH}} = -\frac{\varepsilon^2 \varkappa}{3D}\sum_i N_i z_i^2 = -\frac{kTV\varkappa^3}{12\pi}, \tag{4-27}$$

$$p^{\text{DH}} = -\frac{\varepsilon^2 \varkappa}{6D}\sum_i n_i z_i^2 = -\frac{kT\varkappa^3}{24\pi}. \tag{4-28}$$

Other useful thermodynamic properties are

$$\mu_j^{\text{DH}} = \left.\frac{\partial F^{\text{DH}}}{\partial N_j}\right)_{T, V, N_i} = -\frac{\varepsilon^2 z_j^2 \varkappa}{2D}, \tag{4-29}$$

$$\ln \gamma_j^{\text{DH}} = \frac{\mu_j}{kT} = -\frac{\varepsilon^2 z_j^2 \varkappa}{2DkT} \tag{4-30}$$

where $\mu$ is the chemical potential and $\gamma$ is the activity coefficient.

The preceding expressions for the electrical contribution to the thermodynamic properties are to be added to the corresponding ideal gas expressions. It is noted that the Debye-Hückel contributions are negative, thus describing a net attractive force between the charged particles.

The equilibrium concentration equation including electrical effects will, of course, also differ from the perfect gas expression. For the general ionization reaction

$$A_i \rightleftharpoons A_{i+1} + e - I_i \tag{4-31}$$

the thermodynamic equilibrium condition is

$$\mu_i = \mu_{i+1} + \mu_e. \tag{4-32}$$

Noting that $\mu = \mu^{\text{ideal}} + \mu^{\text{corr}}$, the equilibrium concentration equation is found to be

$$\frac{N_{i+1} N_e}{N_i} = \frac{Z_{i+1} Z_e}{Z_i} \exp\left[-\frac{1}{kT}\left(I_i + \mu_{i+1}^{\text{corr}} + \mu_e^{\text{corr}} - \mu_i^{\text{corr}}\right)\right] \tag{4-33}$$

which can also be written as

$$\frac{N_{i+1} N_e}{N_i} = \frac{Z_{i+1} Z_e}{Z_i} \frac{\gamma_i^{\text{corr}}}{\gamma_{i+1}^{\text{corr}} \gamma_e^{\text{corr}}} \exp\left(-\frac{I_i}{kT}\right) \tag{4-34}$$

where the $Z$'s are the perfect gas partition functions as in Chapter 3. Using the value of $\mu$ or $\gamma$ from the Debye-Hückel limiting laws, Eq. (4–29) or (4–30), the equilibrium concentration becomes

$$\frac{N_{i+1} N_e}{N_i} = \frac{Z_{i+1} Z_e}{Z_i} \exp\left\{\frac{-1}{kT}\left[I_i - (z_{i+1}^2 + z_e^2 - z_i^2)\frac{\varepsilon^2 \varkappa}{2D}\right]\right\}. \tag{4-35}$$

It is noted that since $(z_{i+1}^2 + z_e^2 - z_i^2)$ is always positive, the Debye-Hückel limiting law contribution to the equilibrium equation shows up as a decrease in the ionization potential. An apparent decrease in ionization potential has actually been observed in ionized gases (see ref. 5).

## 4.4 Evaluation of the Theory

Since the Debye-Hückel theory was published, a very large number of papers have appeared in which the theory was discussed and criticized and in which improvements, both of a theoretical and semiempirical nature, were suggested. Since a number of assumptions were involved in the formulation of the theory, the conditions under which it is valid are not at all obvious.

Fowler (ref. 6) has shown that the Poisson-Boltzmann equation, (4–5), is valid only if certain fluctuation terms are small enough to be neglected. In order for any solution of the Poisson-Boltzmann equation to be valid, it can be shown (see, e.g., Fowler and Guggenheim (ref. 3)) that two checks for the self-consistency of the solution must hold, namely,

$$\phi_i/z_i = \phi_j/z_j, \tag{4–36}$$

$$\partial\phi_i/\partial z_j = \partial\phi_j/\partial z_i. \tag{4–37}$$

The first condition states that the potential is directly proportional to the charge. It is noted that, while the Poisson-Boltzmann equation, (4–5), does not satisfy this condition, the linearized equation, (4–9), does. It can be determined whether the second condition is fulfilled by calculating the chemical potential by the Debye charging process and by the Güntelberg charging process (see ref. 3). If the resulting expressions are identical, Eq. (4–37) is satisfied. (The method of calculating the chemical potential, starting from $\phi_j$, which is described in this chapter is equivalent to the Debye charging process.) It can easily be determined that the Debye-Hückel solution satisfies both self-consistency checks.

A number of attempts have been made to improve the Debye-Hückel theory. Among the attempts to accomplish this by purely theoretical means, one method involves the use of a distribution function other than the Boltzmann distribution function. Another method is to solve the Poisson-Boltzmann equation including the higher order terms of the power series expansion of the exponential term. However, all of these attempts have led to solutions which do not satisfy the self-consistency checks, and therefore, it is by no means certain that they are more valuable than the Debye-Hückel solution.

From other considerations, however, it is possible to draw some conclusions as to the validity and applicability of the Debye-Hückel theory. It is found that the limiting laws agree quite closely with experimental data of electrolytes at low concentrations, becoming exact as the concentration approaches zero. Furthermore, when the expressions with the finite distance of closest approach are used, agreement with experimental data is obtained at somewhat higher concentrations. The method here is to choose the value of the distance of closest approach which gives the best agreement. The validity of the Debye-Hückel theory for electrolytes of very low concentration is therefore well established experimentally.

The same conclusions can be drawn from theoretical considerations. Kirkwood and Poirier (ref. 7) concluded that, for ions of zero size, the Poisson-Boltzmann equation is valid for the coefficient of the first power term in an expansion of a more exact expression. Montroll and Ward (ref. 8), who developed a quantum statistical theory of interacting particles, found that, for an electron gas, their theory reduced to the Debye-Hückel theory in the classical (high temperature, low density) limit. From some of the more exact theories of electrolytes, the conclusion is the same.

That the Debye-Hückel theory is valid for low charged particle densities has therefore been established, both experimentally and theoretically. In the application of the theory, however, it is important to know the maximum density for which the theory is valid. This point will be discussed in Section 4.7.

Further details and discussion of the Debye-Hückel theory can be found in texts by Fowler (ref. 6), Fowler and Guggenheim (ref. 3), Harned and Owen (ref. 9), Robinson and Stokes (ref. 4),

Landau and Lifshitz (ref. 10), and Hill (ref. 11). Many additional references are given in these works.

## 4.5 The Debye Length

It is of interest to note the properties of the quantity $\varkappa$ given by Eq. (4–8). $\varkappa$ has the dimensions of inverse length and so it is natural to speak of its reciprocal:

$$h = \frac{1}{\varkappa} = \left(\frac{DkT}{4\pi\varepsilon^2 \Sigma_i n_i z_i^2}\right)^{1/2}. \tag{4–38}$$

The length $h$ is variously called the Debye length, the Debye-Hückel radius, the Debye shielding distance, or the mean thickness of the ionic atmosphere. Various physical interpretations of this length have been proposed. Some of the most common will be summarized here.

Debye and Hückel stated that the length $h$ is a measure of the thickness of the ionic atmosphere. Fowler and Guggenheim (ref. 3) showed that the maximum value of $dq/dr$, where $q$ is the charge, occurs at $r = h$. Spitzer (ref. 12) states that $h$ is a measure of the distance over which the electron charge density can deviate appreciably from the ion charge density. If $h$ is small compared with other lengths of interest, such as the mean free path, an ionized gas is called a plasma. Spitzer also notes that when a solid surface is in contact with a plasma, the thickness of the sheath about the surface is approximately equal to the Debye distance. Kantrowitz and Petschek (ref. 13) state that the Debye distance "is the distance by which the positive and negative charges in a gas will be separated if the thermal energy of the gas is used for charge separation." They also state that $h$ is "the distance in which the electric field of a point charge is shielded by a local increase of concentration of charges of opposite sign." It is noted that the various interpretations of $h$ are essentially the same.

It is possible to obtain an analytical interpretation of the Debye length by starting with the linearized Poisson-Boltzmann equation, (4–9). This equation can be nondimensionalized by defining the dimensionless quantities $r' = r/L$ and $\phi' = \phi/\phi_c$ where $L$ is some characteristic length and $\phi_c$ is a characteristic potential. Substituting for $r$ and $\phi$ in Eq. (4–9) leads to

$$\nabla^2 \phi' = L^2 \varkappa^2 \phi' = \frac{L^2}{h^2} \phi'. \tag{4–39}$$

Clearly, the dimensionless quantity $L^2/h^2$ is an important parameter in the Debye-Hückel theory. For a singly ionized gas, $n_e z_e^2 = n_1 z_1^2$, and the Debye distance is

$$h = \left( \frac{DkT}{8\pi\varepsilon^2 n_e} \right)^{1/2}. \tag{4–40}$$

Therefore

$$\frac{L^2}{h^2} = L^2 \left( \frac{8\pi\varepsilon^2 n_e}{DkT} \right). \tag{4–41}$$

A logical choice for $L$ would be the average distance between charged particles, in this case, $L = \bar{r} = (2n_e)^{-1/3}$. Substituting this in Eq. (4–41) leads to

$$\frac{L^2}{h^2} = \frac{\bar{r}^2}{h^2} = 4\pi \frac{\varepsilon^2}{D(2n_e)^{-1/3} kT}$$

$$= 4\pi \frac{\varepsilon^2}{\bar{r} DkT}. \tag{4–42}$$

Now $\varepsilon^2/D\bar{r}$ is the Coulomb potential energy of two charged particles at a distance $\bar{r}$ apart, and $kT$ is the thermal energy of these particles. Therefore, the dimensionless parameter is roughly equal to the ratio of the Coulomb energy to the thermal energy, or

$$\frac{\bar{r}^2}{h^2} = 4\pi \frac{E^{\text{coul}}}{E^{\text{ther}}}. \tag{4–43}$$

The Debye length is therefore

$$h^2 = \frac{\bar{r}^2}{4\pi} \frac{E^{\text{ther}}}{E^{\text{coul}}} \tag{4–44a}$$

or

$$h = \bar{r} \left( \frac{1}{4\pi} \frac{E^{\text{ther}}}{E^{\text{coul}}} \right)^{1/2}. \tag{4–44b}$$

It would be logical to assume that if the thermal energy of an ionized gas is an order of magnitude, say $4\pi$, larger than the Coulomb energy, it may not be necessary to consider electrical deviations from the perfect gas law. For this case, with $E^{\text{ther}} = 4\pi E^{\text{coul}}$, $h = \bar{r}$. Therefore, the Debye-Hückel corrections appear to be important only when the Debye distance is less than the mean distance between charged particles.

Equation (4–43) was derived for the case of single ionization only. For the case of a gas containing electrons and only one other type of positive ion, e.g., doubly ionized atoms or triply ionized atoms, the result is the same. For gases containing more than one type of positive ion, the constant would differ somewhat from $4\pi$, but the general conclusion would remain valid.

It should be noted that the Debye length is sometimes given (see for example Delcroix, ref. 14) as

$$h = \left( \frac{DkT}{4\pi\varepsilon^2 n_e} \right)^{1/2}. \tag{4–45}$$

This result can be obtained by assuming that, while the electrons obey the Boltzmann distribution function, Eq. (4–4), the density of the ions is uniform, i.e., they can be considered stationary in space. It can be shown that the corresponding Debye length is given by Eq. (4–45) and that this expression is valid for both singly and multiply ionized gases. Although Eq. (4–45) is approximate, the error resulting from the use of this simpler expression for the Debye length will, in most cases, be small.

## 4.6 Magnitude of the Debye-Hückel Correction Term

It is useful to know when the magnitude of the corrections given by the Debye-Hückel theory become large enough so that they must be considered in calculations. That is, it is useful to determine the maximum charged particle density for which the perfect gas expressions are valid. In order to illustrate the magnitude of these

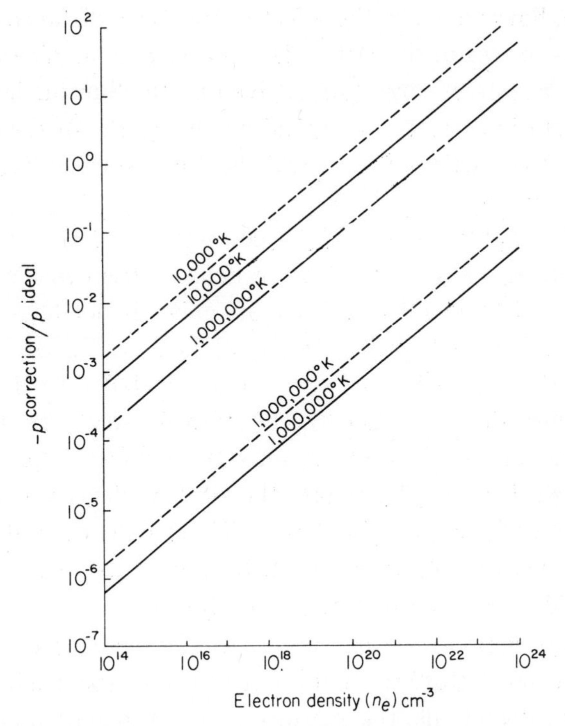


FIG. 4–1. Debye-Hückel and modified Debye-Hückel pressure corrections. Limiting law: single ionization, —; double ionization, — — —; fully ionized nitrogen, — ———.

corrections, the ratio of the Debye-Hückel limiting law pressure correction, Eq. (4–28), to the perfect gas pressure is plotted in Fig. (4–1) as a function of electron density at constant temperature. This pressure ratio is given by

$$\frac{p^{\mathrm{DH}}}{p^{\mathrm{ideal}}} = -\frac{\varkappa^3}{24\pi \Sigma_s n_s}. \tag{4-46}$$

The pressure corrections are plotted at 10,000° and 1,000,000 °K, since it is assumed that these are the highest and lowest temperatures of interest. Therefore, the correction will lie somewhere between them. The pressure correction for a finite distance of closest approach, Eq. (4–25), is not plotted in Fig. 4–1. It is noted here, however, that the effect of the finite distance of closest approach is to reduce the Debye-Hückel correction. Therefore, the maximum (negative) correction is given by the limiting law. Since the electrostatic corrections depend on the charge of the particles, but not on the particular gas involved, the results of Fig. 4–1 are applicable for any gas.

Several simplifications were necessary in order to plot these equations in Fig. 4–1. The electron density was considered to be independent of temperature. This, of course, is not true in reality since they are related by Eq. (4–33). However, Eq. (4–35) shows that when the Debye-Hückel correction is included in the equilibrium concentration equation, the electron density is no longer an explicit function of temperature, since the exponential term is also a function of density. Therefore, the electron density would have to be calculated by trial and error. This would be quite tedious especially since the electronic partition functions consist of series which would have to be summed for each trial.

Since the composition of the gas was not calculated from the equilibrium concentration equation, it had to be specified. In order to simplify calculations, the gas was assumed to consist entirely of electrons plus one type of ion, i.e., singly charged ions, doubly charged ions, etc. Thus, the gas might be called fully singly ionized, fully doubly ionized, and so on. Actually, the results plotted for this type of gas are quite meaningful since the effect of the charge of the ions is clearly shown. Also, since there are no neutral particles in this gas, Eq. (4–46) represents the ratio of electrostatic pressure of the charged particles to the perfect gas pressure of these same charged particles. If the perfect gas pressure included the pressure due to neutral particles, the meaning of the results would be somewhat less clear.

To help clarify the preceding remarks, a sample calculation will be given. Equation (4–46) can be written

$$\frac{p^{\mathrm{DH}}}{p^{\mathrm{ideal}}} = -\frac{\pi^{1/2}\,\varepsilon^3(\Sigma_i\, n_i z_i^2)^{3/2}}{3D^{3/2}(kT)^{3/2}\,\Sigma_s n_s}. \tag{4–47}$$

For the case of a fully singly ionized gas, only singly charged ions and electrons exist in the gas, and since the gas is electrically neutral, $n_1 = n_e$. Therefore, the summation over all charged particles in the numerator of Eq. (4–47) is equal to $2n_e$, and the summation over all particles, both charged and neutral, in the denominator is also equal to $2n_e$. Therefore, Eq. (4–47) can be written for a fully singly ionized gas as

$$\frac{p^{\mathrm{DH}}}{p^{\mathrm{ideal}}} = -\frac{(2\pi)^{1/2}\,\varepsilon^3\, n_e^{1/2}}{3D^{3/2}(kT)^{3/2}} = -5.7\times 10^{-5}\,\frac{n_e^{1/2}}{T^{3/2}}. \tag{4–48}$$

The substitution of various values of electron density in Eq. (4–48) for a constant temperature results in the limiting law lines in Fig. 4–1.

Because of the simplifications involved, the results presented in Fig. 4–1 must be interpreted very carefully. For example, Fig. 4–1 gives the pressure correction for a gas having an electron density of $10^{22}$ cm$^{-3}$ at 10,000 °K. However, calculations based on the equilibrium concentration equation, (4–34), show that electron densities of this magnitude do not exist in equilibrium at a temperature of only 10,000 °K. Also, at 1,000,000 °K, a gas would consist largely of multiply ionized atoms unless, of course, the gas was hydrogen. In other words, a fully singly ionized gas at 1,000,000 °K is not possible for gases of higher atomic number. Figure 4–1 is valuable in that it compares the magnitude of the electrostatic pressure corrections and the perfect gas pressure, thus showing if these corrections are significant.

Equation (4–46) is plotted in Fig. 4–1 for the case of a fully singly ionized gas, a fully doubly ionized gas at temperatures of 10,000° and 1,000,000 °K, and for the case of a fully septuply ionized gas at 1,000,000 °K which could represent, for example, fully ionized nitrogen. Above the horizontal line at $p^{\text{corr}}/p^{\text{ideal}} = 10^{-2}$, the pressure corrections are over 1% of the perfect gas pressure, and consequently can be considered significant. It is seen that, in general, the pressure corrections are not significant below an electron density of about $10^{16}$ cm$^{-3}$. Since the electron density does not normally exceed $10^{16}$ cm$^{-3}$ in most aerodynamic problems, the perfect gas assumption is usually valid for these applications.

## *4.7 Validity of the Debye-Hückel Theory at Higher Densities*

### 4.7.1 Fowler-Berlin-Montroll Criterion

Berlin and Montroll (ref. 15) showed that the Debye-Hückel limiting law expressions were theoretically valid up to a critical or limiting concentration. Their critical concentration can be obtained from

$$\sum_i n_i = \frac{1}{8\pi} \varkappa^3 \tag{4–49}$$

For a gas containing only singly charged ions and electrons, the limiting electron density is given by

$$n_e^{\text{lim}} = \frac{1}{2\pi} \left( \frac{DkT}{\varepsilon^2} \right)^3 . \tag{4–50}$$

This limiting electron density is plotted in Fig. 4–2. According to this criterion, the Debye-Hückel limiting law is valid for all points below this curve, but is not accurate above the curve.

Fowler (ref. 6), in considering the legitimacy of ignoring fluctuation terms in the derivation of Poisson's equations, concluded that these terms could be neglected if

$$\frac{|z_i z_j| \varepsilon^2 \varkappa}{2DkT} \ll 1. \tag{4–51}$$

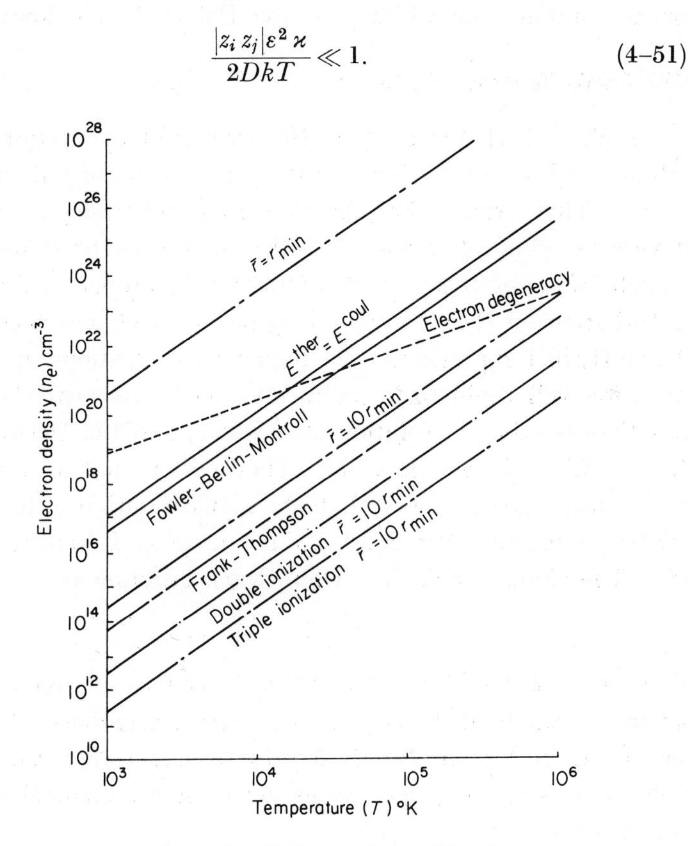


FIG. 4–2. Regions of validity of Debye-Hückel theory.

A limiting density can be found by equating both sides of Eq. (4–51). For a gas containing only singly charged ions and electrons, the limiting electron density is identical to that given by the Berlin-Montroll expression, Eq. (4–50).

For any gas consisting only of electrons and a single species of ions of valence $z$, the Berlin-Montroll expression and the Fowler expression give identical results, provided $|z_i z_j|$ is taken as the valence of the electron times the valence of the ion. From theoretical considerations, therefore, Eq. (4–49) or its equivalent expression for multiple ionization is well established as a criterion for determining the validity of the Debye-Hückel limiting law.

### 4.7.2 Frank-Thompson Criterion

Frank and Thompson (ref. 16) arrived at a criterion for the validity of the Debye-Hückel theory by means of inductive reasoning. They argued that the Debye-Hückel theory is successful at very low concentrations, since the nearest-neighbor ions are far enough apart for their direct interaction to produce only a small perturbation on the dominant long-range effect described by the Debye-Hückel expressions. At higher concentrations, however, a given ion will begin to be more affected by its nearest-neighbor ions than by the ion solution as a whole, and the Debye-Hückel theory will no longer be valid. They reason that an ion should see its image farther away than the actual location of its nearest-neighbor ions, and that the image is a distance $h$ from the central ion. Therefore, Frank and Thompson's criterion is

$$h = 1/\varkappa > \bar{r} \tag{4–52}$$

where $\bar{r} = (\Sigma_i n_i)^{-1/3}$ is the average distance between charged particles, i.e., the distance between nearest neighbors. Replacing the inequality sign in Eq. (4–52) by an equal sign, the limiting electron density for a gas consisting of singly charged ions and electrons is

$$\begin{aligned} n_e^{\text{lim}} &= \frac{1}{2}\frac{1}{(4\pi)^3}\left(\frac{DkT}{\varepsilon^2}\right)^3 \\ &= 2.5 \times 10^{-4}\left(\frac{DkT}{\varepsilon^2}\right)^3. \end{aligned} \tag{4–53}$$

Equation (4–53) is also plotted in Fig. 4–2. It is noted that when $E^{\text{ther}} = 4\pi E^{\text{coul}}$, Eq. (4–44b) reduces to the Frank-Thompson criterion.

### 4.7.3 Criterion Based on the Equivalence of Thermal and Coulomb Energies

Instead of the Frank-Thompson criterion which states, in effect, that the thermal energy must be an order of magnitude greater than the Coulomb energy, it might be argued that the Debye-Hückel theory is valid as long as the Coulomb energy does not exceed the thermal energy. The criterion is, therefore, $E^{\text{ther}} = E^{\text{coul}}$, or

$$kT = \frac{|z_i z_j| \varepsilon^2}{D\bar{r}}. \tag{4–54}$$

The average separation of charged particles must therefore not be less than

$$\bar{r} = \left(\sum_i n_i\right)^{-1/3} = \frac{|z_i z_j| \varepsilon^2}{DkT}. \tag{4–55}$$

For a gas containing only singly charged ions and electrons, the limiting electron density is

$$n_e^{\lim} = \frac{1}{2}\left(\frac{DkT}{\varepsilon^2}\right)^3. \tag{4–56}$$

Equation (4–56) is also plotted in Fig. 4–2. It is observed that this criterion is identical to the Fowler-Berlin-Montroll criterion [Eq. (4–49)] except for the factor $1/\pi$. Equation (4–49) states, in effect, that $E^{\text{ther}} = \pi^{1/3} E^{\text{coul}} = 1.46\, E^{\text{coul}}$, and thus the two criteria are very nearly the same. This criterion, then, gives a physical explanation of why the Fowler-Berlin-Montroll criterion is logical.

Another reason why Eq. (4–49) or (4–55) represents accurate criteria for the validity of the Debye-Hückel theory can be found by examining the Debye-Hückel expressions themselves. By combining Eqs. (4–4), (4–6), and (4–14), the number density can be expressed as

$$n_i'(r) = n_i \left[1 - \frac{z_i z_j \varepsilon^2}{DkT} \frac{e^{\varkappa a}}{1 + \varkappa a} \frac{e^{-\varkappa r}}{r}\right] \tag{4–57}$$

which can be written as

$$n_i'(r) = n_i\left[1 - \frac{z_i z_j}{|z_i z_j|}\left(\frac{|z_i z_j|\varepsilon^2}{DkT}\frac{e^{\varkappa a}}{1+\varkappa a}\frac{e^{-\varkappa r}}{r}\right)\right]. \tag{4–58}$$

Now as $r$ approaches zero, the quantity in parentheses approaches infinity. Thus, the number density approaches either plus or minus infinity depending on whether $z_i$ and $z_j$ represent the valences of like charges or unlike charges. However, it is physically meaningless for $n'(r)$ to be infinite or to have any negative value. Therefore, it is necessary that

$$\frac{|z_i z_j|\varepsilon^2}{DkT}\frac{e^{\varkappa a}}{1+\varkappa a}\frac{e^{-\varkappa r}}{r} \leqslant 1. \tag{4–59}$$

Before proceeding, it should be noted that the Debye-Hückel expressions were derived for the condition that

$$\frac{z_i \varepsilon \phi_j}{kT} < 1. \tag{4–60}$$

Substituting the potential [Eq. (4–14)] into Eq. (4–60) leads to Eq. (4–59). Therefore, Eq. (4–59) represents a necessary condition for the validity of the Debye-Hückel theory.

Now the maximum value of Eq. (4–59) will occur at the minimum value of $r$, namely, $a$. Therefore, substituting $r = a$ in Eq. (4–59) leads to

$$\frac{|z_i z_j|\varepsilon^2}{DkT}\frac{1}{a(1+\varkappa a)} \leqslant 1. \tag{4–61}$$

The distance of closest approach is now equated to the minimum average distance between charged particles as given by Eq. (4–55):

$$a = \bar{r}_{\min} = \frac{|z_i z_j|\varepsilon^2}{DkT} = A. \tag{4–62}$$

Substituting Eq. (4–62) into Eq. (4–61) results in

$$\frac{1}{1+\varkappa A} \leqslant 1. \tag{4–63}$$

Since $\varkappa$ and $A$ are always positive, the above inequality is always true.

Thus, it is seen that criterion based on the equivalence of thermal and Coulomb energies leads to an expression for the distance of closest approach which is consistent with the Debye-Hückel theory. Furthermore, this expression is physically reasonable. According to the theory of recombination of ions (see for example Loeb, ref. 17), one of the conditions necessary for the recombination of oppositely charged particles to occur is that the Coulomb potential energy must be greater than the thermal energy. Consequently, particles of like sign will, on the average, not come closer together than $A$ due to their mutual repulsive Coulomb energy, and particles of unlike sign which are separated by distance less than $A$ are likely to recombine. The distance $A$ is also essentially equivalent to the impact parameter which is used in the theory of collisions between charged particles.

A similar value of the distance of closest approach has been used to predict the properties of electrolytes. It can be shown (see for example ref. 9) that the probability of an $i$ ion being at a distance $r$ from an oppositely charged $j$ ion possesses a minimum value, the distance $d$, such that

$$d = \frac{|z_i z_j|\varepsilon^2}{2DkT} = \frac{A}{2}. \tag{4–64}$$

Assuming that all ions of unlike sign closer together than $d$ are associated, Bjerrum (ref. 18) was able to formulate a theory of ionic association for electrolytes. Using the appropriate value of $d$ as the distance of closest approach in the Debye-Hückel expressions, together with his calculated value of the degree of association, Bjerrum was able to obtain improved agreement between theory and experiment in some cases. However, it should be noted that if $d$ is substituted for $a$ in Eq. (4–61) the result is

$$\frac{2}{1 + \varkappa d} \leqslant 1. \tag{4–65}$$

This inequality is only satisfied when $\varkappa d \geqslant 1$ which would not be the case at low concentrations or high temperatures. Therefore,

Eq. (4–62) appears to give a more reasonable value of the distance of closest approach to be used with the Debye-Hückel theory.

It is important to remember that the foregoing discussion is not theoretically accurate, since the right-hand side of Eq. (4–54) gives the Coulomb potential energy between two charged particles only in the absence of all other charged particles. In other words, the shielding effect of other ions and electrons is neglected. However, it should be a good approximation at low charged particle densities.

### 4.7.4 Criterion Based on the Minimum Distance of Closest Approach

While the value of the distance of closest approach given by Eq. (4–62) does satisfy Eq. (4–61), it is of interest to determine the minimum value of $a$ which will satisfy this condition. Therefore, replacing the inequality sign by an equal sign, the condition becomes

$$\frac{A}{a_{\min}(1 + a_{\min}\,\varkappa)} = 1. \tag{4–66}$$

The solution to this equation is

$$a_{\min} = -\frac{1}{2\varkappa} + \left(\frac{1}{4\varkappa^2} + \frac{A}{\varkappa}\right)^{1/2} \tag{4–67}$$

where $a_{\min}$ is the smallest value which is consistent with Debye-Hückel theory.

The same result can be obtained for the case of point charges, that is, letting $a = 0$ in Eq. (4–59) leads to

$$\frac{|z_i\, z_j|\varepsilon^2}{DkT}\,\frac{e^{-\varkappa r}}{r} \leqslant 1. \tag{4–68}$$

Since the minimum value of $r$ is sought, $\varkappa r$ will be small and the exponential term can be replaced by the first two terms of its series expansion. The result is

$$\frac{A}{r_{\min}(1 + r_{\min}\,\varkappa)} = 1. \tag{4–69}$$

Equation (4–69) is equivalent to Eq. (4–66) and the solution is again (letting $1/\varkappa = h$)

$$r_{\min} = -\tfrac{1}{2}h + (\tfrac{1}{4}h^2 + Ah)^{1/2} \tag{4–70}$$

where $r_{\min}$ is the minimum average separation between charged particles which is consistent with the Debye-Hückel theory.

A necessary condition for the valid application of the Debye-Hückel expressions is obviously that the average distance between charged particles must not be less than $r_{\min}$, that is,

$$\left(\sum_i n_i\right)^{-1/3} \geqslant -\tfrac{1}{2}h + (\tfrac{1}{4}h^2 + Ah)^{1/2}. \tag{4–71}$$

For a gas containing only singly charged ions and electrons, the limiting electron density can be found from the following equation:

$$(2n_e)^{-1/3} = -\frac{1}{2}\sqrt{\frac{DkT}{8\pi\varepsilon^2 n_e}} + \left(\frac{1}{4}\frac{DkT}{8\pi\varepsilon^2 n_e} + A\sqrt{\frac{DkT}{8\pi\varepsilon^2 n_e}}\right)^{1/2}. \tag{4–72}$$

The solution of Eq. (4–72) is found to be

$$n_e^{\lim} = 1.53 \times 10^3 \left(\frac{DkT}{\varepsilon^2}\right)^3. \tag{4–73}$$

Equation (4–73) is plotted in Fig. 4–2. It is seen that this criterion predicts that the Debye-Hückel theory will be valid at electron densities of the order of $10^3$ times greater than the densities predicted by Eq. (4–49) or (4–56). The reason for this somewhat unreasonable value is readily seen. The condition given by Eq. (4–70) states that no particles should be closer together than $r_{\min}$, while Eq. (4–71) states that, on the average, no particles will be closer together than $r_{\min}$. Therefore, many particles will be separated by less than $r_{\min}$.

To assure that the number of particles closer together than $r_{\min}$ is small, a reasonable criterion would be $\bar{r} \geqslant 10 r_{\min}$, or

$$\left(\sum_i n_i\right)^{-1/3} \geqslant 10\,[-\tfrac{1}{2}h + (\tfrac{1}{2}h^2 + Ah)^{1/2}]. \tag{4–74}$$

For a gas containing only singly charged ions and electrons, the limiting electron density is given by

$$n_e^{\lim} = 10^{-3}\left(\frac{DkT}{\varepsilon^2}\right)^3. \tag{4–75}$$

It is seen that this limiting electron density is about $10^6$ times smaller than that given by Eq. (4–73), and is of the same order of magnitude as that given by the Frank-Thompson criterion. The Frank-Thompson criterion is based upon the concept that the image of an ion is a distance $h$ away. This concept is rather nebulous since it was seen previously that the physical meaning of the Debye length is somewhat vague. However, the criterion given by Eq. (4–74) is physically reasonable, and since it agrees closely with the Frank-Thompson criterion, the Frank-Thompson criterion is also reasonable.

For a gas containing only doubly charged ions and electrons and letting $z_i = z_j = 2$ in $A$, the limiting electron density, as given by Eq. (4–74), is

$$n_e^{\lim} = 1.4 \times 10^{-5}\left(\frac{DkT}{\varepsilon^2}\right)^3. \tag{4–76}$$

For a gas containing only triply ionized atoms and electrons, and letting $z_i = z_j = 3$ in $A$, Eq. (4–74) leads to

$$n_e^{\lim} = 1.25 \times 10^{-6}\left(\frac{DkT}{\varepsilon^2}\right)^3. \tag{4–77}$$

Equations (4–76) and (4–77) are plotted in Fig. 4–2. The Frank-Thompson criterion for double and triple ionization leads to somewhat higher values of the limiting electron density.

### 4.7.5 Electron Degeneracy Criterion

All of the formulae presented thus far are valid only if classical (Boltzmann) statistics are applicable. At high densities, it may be necessary to use formulas based on quantum statistics, or to apply quantum corrections to the classical expressions. A gas for which quantum statistics must be used is said to be degenerate. It is therefore important to know when the error resulting from the use of Boltzmann statistics is appreciable. Only a criterion for the validity of Boltzmann statistics will be discussed here. A discussion of quantum corrections can be found in ref. 19.

From a comparison of the distribution functions obtained from quantum and Boltzmann statistics, it is found (see for example ref. 3) that if

$$\frac{nh^3}{(2\pi mkT)^{3/2}} \ll 1, \tag{4-78}$$

the Boltzmann statistics are valid. In the above expression, $m$ is the mass of the particle under consideration and $h$ is Planck's constant. Since the left-hand side of Eq. (4–78) is an inverse function of mass, its largest values will occur for the case of electrons. Since all other particles are much heavier than electrons, it is rarely necessary to consider the degeneracy of heavier particles.

For electrons, Eq. (4–78) becomes

$$n_e \ll \left(\frac{2\pi m\, kT}{h^2}\right)^{3/2} = 2.4 \times 10^{15}\, T^{3/2}. \tag{4-79}$$

In order to calculate the limiting electron density, it will be assumed that if $n_e$ does not exceed $\frac{1}{10}$ of the right-hand side of Eq. (4–79), electron degeneracy is not important, i.e.,

$$n_e^{\text{lim}} = \tfrac{1}{10}\,(2.4 \times 10^{15})\, T^{3/2}. \tag{4-80}$$

Equation (4–80) is plotted in Fig. 4–2.

### 4.7.6 Comparison of Criteria

It is noted that all of the criteria discussed in this chapter, with the exception of the electron degeneracy criterion, had the same analytical form for a gas consisting of electrons and only one species of ions, namely,

$$n_e^{\lim} = (\text{const})\left(\frac{DkT}{\varepsilon^2}\right)^3 . \tag{4-81}$$

Consequently, in Fig. 4–2, which is a logarithmic plot, all of these criteria are represented by straight lines of identical slopes. The area below a particular line represents the region in which the corresponding criterion predicts the Debye-Hückel expressions are valid. For example, the area below the Frank-Thompson line represents the region in which this criterion predicts the Debye-Hückel theory is valid for a singly ionized gas.

The criterion predicting the highest density for a given temperature is that given by Eq. (4–70). However, it was explained in Section 4.7.4 that this criterion is not very meaningful. It is significant only in that the Debye-Hückel expressions could be expected to lead to grossly inaccurate results if applied in the region above the line.

Lower on the graph, the Fowler-Berlin-Montroll criterion and the criterion based on the equivalence of thermal and Coulomb energies are seen to be nearly the same. The latter is significant in that it illustrates the physical meaning of the Fowler-Berlin-Montroll criterion. The investigations of Fowler, Berlin, and Montroll have shown that their criterion represents the theoretical limits of validity of the Debye-Hückel theory. Furthermore, the energy equivalence criteria states that the average distance between charged particles must not be less the quantity $A$ which is a reasonable approximation for the distance of closest approach. Since, for the region above this curve, $\bar{r} < A$, short-range forces can be expected to be important.

Lowest on the graph are the Frank-Thompson criterion and the criterion for $\bar{r} \geqslant 10 r_{\min}$. These criteria are nearly identical and it is believed that they represent the practical limit of the Debye-Hückel expressions; that is, for densities lower than the limiting density predicted by these criteria, the Debye-Hückel expressions can be expected to give accurate results. For higher densities the accuracy can be expected to drop off. This conclusion is supported by experimental data for the properties of electrolyte solutions. The Debye-Hückel limiting law is found to represent data for electrolytes, consisting of only singly charged ions (1–1 electrolytes), quite accurately up to concentrations of about 0.001 M which corresponds to a charged particle density of about $6 \times 10^{17}$ cm$^3$ (see for example ref. 4). For $T = 300$ °K, and $D = 78$ (the value for water), Frank and Thompson's criteria, Eq. (4–53), gives the limiting density as $7 \times 10^{17}$ cm$^{-3}$. Thus, for electrolytes at least, Frank and Thompson's criterion gives good results.

Figure 4–2 shows that at temperatures of about 100,000 °K, the electron degeneracy criterion limits the validity of the Debye-Hückel expressions, and at somewhat over 1,000,000 °K it replaces the Frank-Thompson criterion. In the region above the degeneracy line, quantum corrections must be added to the classical expressions.

In conclusion, the Debye-Hückel theory can be applied with confidence below the Frank-Thompson limiting density. Between this density and the Fowler-Berlin-Montroll limiting density, the theory is valid in principle but its accuracy may decrease. Also, at higher temperatures, the limiting density may be given by the electron degeneracy criterion rather than the Fowler-Berlin-Montroll criterion.

It was mentioned previously that when $E^{\text{ther}} = 4\pi E^{\text{coul}}$, that is, when the Frank-Thompson relation holds, it may not be necessary to consider the Debye-Hückel corrections. This point can be investigated by considering the Debye-Hückel pressure correction. The ratio of the Debye-Hückel pressure, Eq. (4–28), to the perfect gas pressure is

$$\frac{p^{\text{DH}}}{p^{\text{ideal}}} = -\frac{\varkappa^3}{24\pi \Sigma_s n_s}. \tag{4–82}$$

Since the Frank-Thompson criterion is $(\Sigma_i n_i)^{-1/3} = \varkappa^{-1}$, Eq. (4–82) becomes

$$\frac{p^{\text{DH}}}{p^{\text{ideal}}} = -\frac{1}{24\pi} \frac{\Sigma_i n_i}{\Sigma_s n_s}. \tag{4–83}$$

For a gas consisting only of charged particles, the two summations are equal, and

$$\frac{p^{\mathrm{DH}}}{p^{\mathrm{ideal}}} = -\frac{1}{24\pi} = -0.0133. \tag{4–84}$$

Therefore, if the Debye-Hückel pressure correction is used in the region in which its validity is certain, its magnitude will never exceed 1.33% of the perfect gas pressure. If the criterion based on $\bar{r} \geqslant 10r_{\min}$ is used instead, the magnitude of the pressure correction will never exceed 2.63% of the perfect gas pressure. Consequently, it is seen that if the Debye-Hückel corrections are used where their validity is certain, their magnitude is probably negligibly small, and if the corrections are used where their magnitude is appreciable, their accuracy is probably in doubt. The necessity of theories which will predict thermodynamic properties beyond the densities for which the Debye-Hückel theory is valid is apparent. It is also seen that electrostatic corrections are, in general, important only when the Debye length is less than the average distance between charged particles.

## 4.8 List of Symbols

Latin symbols:

$a$ — distance of closest approach

$a_{\min}$ — minimum distance of closest approach for Debye-Hückel theory

$A$ — expression for the distance of closest approach

$C'$ — constant of integration

$C''$ — constant of integration

$d$ — expression for the distance of closest approach $(A/2)$

$D$ — dielectric constant

$E$ — internal energy

$F$ — Helmholtz free energy

$h$ — Debye length $(1/\varkappa)$

$I$ — ionization energy

$k$ — Boltzmann's constant
$L$ — characteristic length
$n$ — average number density of particles $(N/V)$
$n'(r)$ — nonaverage number density
$N$ — number of particles
$p$ — pressure
$q$ — charge
$r$ — spherical distance from particle
$r'$ — dimensionless distance $(r/L)$
$\bar{r}$ — average distance between charged particles
$r_{\min}$ — minimum average distance between charged particles for Debye-Hückel theory
$T$ — absolute temperature
$V$ — volume
$W$ — potential energy of ion
$z$ — valence of charged particle
$Z$ — partition function

Greek symbols:

$\gamma$ — activity coefficient
$\varepsilon$ — electronic charge
$\varkappa$ — inverse Debye length
$\mu$ — chemical potential
$\rho$ — charge density
$\phi$ — electrical potential
$\phi'$ — dimensionless electrical potential
$\phi_c$ — characteristic potential
$\phi^*$ — electrical potential minus self-potential
$s$ — summation index for all gas particles

Subscripts:

$e$ — electron
$i$ — charged particle; summation index for charged particles
$j$ — charged particle
$s$ — summation index for all gas particles

Superscripts:

corr — correction
coul — Coulomb
DH — Debye-Hückel
ideal — ideal
ther — thermal

## References


1. Debye, P., and E. Hückel, "On the Theory of Electrolytes, I. Freezing Point Depression and Related Phenomena." *Physikalische Zeitschrift*, Vol. 24, No. 9, 1923, pp. 185–206. Also: *The Collected Papers of Peter J. W. Debye.* Interscience, New York, 1954, pp. 217–263.
2. Rosseland, S., "Electrical State of a Star." *Monthly Notices of the Royal Astronomical Society*, Vol. 84, No. 9, 1924, pp. 720–728.
3. Fowler, R., and E. A. Guggenheim, *Statistical Thermodynamics.* Cambridge Univ. Press, London and New York, 1956.
4. Robinson, R. A., and R. H. Stokes, *Electrolyte Solutions.* Academic Press, New York, 1955.
5. Olsen, H. N., "Partition Function Cutoff and Lowering of the Ionization Potential in an Argon Plasma." *The Physical Review*, Vol. 124, No. 6, December 15, 1961, pp. 1703–1708.
6. Fowler, R. H., *Statistical Mechanics*, 2nd edition. Cambridge Univ. Press, London and New York, 1936.
7. Kirkwood, J. G., and J. C. Poirier, "The Statistical Mechanical Basis of the Debye-Hückel Theory of Strong Electrolytes." *Journal of Physical Chemistry*, Vol. 58, No. 8, August, 1954, pp. 591–596.
8. Montroll, E. W., and J. C. Ward, "Quantum Statistics of Interacting Particles; General Theory and Some Remarks on Properties of an Electron Gas." *Physics of Fluids*, Vol. 1, No. 1, January–February, 1958, pp. 55–72.
9. Harned, H. S., and B. B. Owen, *The Physical Chemistry of Electrolytic Solutions.* 3rd edition, Reinhold, New York, 1958.
10. Landau, L. D., and E. M. Lifshitz, *Statistical Physics* (translated by E. Peierls and R. F. Peierls). Addison-Wesley, Reading, Massachusetts, 1958.
11. Hill, T. L., *An Introduction to Statistical Thermodynamics.* Addison-Wesley, Reading, Massachusetts, 1960.
12. Spitzer, L., Jr., *Physics of Fully Ionized Gases.* Interscience, New York, 1956.
13. Kantrowitz, A. R., and H. E. Petschek, "An Introductory Discussion of Magnetohydrodynamics," in *Magnetohydrodynamics* (edited by R. K. M. Landshoff). Stanford Univ. Press, Stanford, California, 1957, pp. 3–15.
14. Delcroix, J. L., *Introduction to the Theory of Ionized Gases.* Interscience, New York, 1960.

15. Berlin, T. H., and E. W. Montroll, "On the Free Energy of a Mixture of Ions: An Extention of Kramer's Theory." *Journal of Chemical Physics*, Vol. 20, No. 1, January, 1952, pp. 75–84.
16. Frank, H. S., and P. T. Thompson, "A Point of View on Ion Clouds" in *The Structure of Electrolytic Solutions* (edited by W. J. Hamer). John Wiley, New York, 1959, pp. 113–134.
17. Loeb, L. B., *Basic Processes of Gaseous Electronics*. Univ. of California Press, Berkeley, California, 1955.
18. Bjerrum, N., "Investigation of Ion Association. I." *Det Kgl. Danske Videnskabernes Selskab*, Vol. 7, No. 9, 1926, 48 pp.
19. Duclos, D. P., and A. B. Cambel, "The Equation of State of an Ionized Gas" in *Progress in International Research on Thermodynamic and Transport Properties* (edited by J. F. Masi and D. H. Tsai). Academic Press, New York, 1962, p. 601.

## Chapter 5

# High Pressure Real Gas Effects

## 5.1 Introduction

It was pointed out in Chapter 1 that real gas effects may occur because of (a) high temperatures, (b) high pressures, and (c) high energy radiation. Real gas behavior and effects due to high temperatures are discussed in Chapters 1–4, 6, and 7, whereas the case of high energy collisions is beyond the scope of this monograph. In this chapter the real gas effects due to high pressures are reviewed. Such effects may occur within certain propulsive devices and research facilities.

## 5.2 High Pressure Effects

At low temperatures but high pressures, a gas becomes imperfect both thermally and calorically. This is due to the intermolecular forces and the volume occupied by the molecules. For example, at a temperature of about 300 °K, the deviation from the perfect gas law becomes significant at about 10 atm, while at 1000 °K, the deviation does not become significant until about 100 atm. Although the low temperature-high pressure regime is seldom encountered in gas dynamics, such situations may occasionally arise. Also, the need for highly accurate calculations, such as in the calibration of wind tunnels, may necessitate the consideration of high pressure effects.

The effect of intermolecular forces can be expressed, as a first approximation, in terms of the second virial coefficient, $B(T)$, and the thermal equation of state can be written (ref. 1)

$$\frac{p}{\rho RT} = Z = 1 + B(T)\frac{p}{RT}, \tag{5-1}$$

and thus the correction term, $B(T)$, is a function of the particular gas. However, for the most common gases, it is possible to introduce dimensionless variables and rewrite the equation in universal form. This method is based on van der Waals' law of corresponding states. Reduced properties, $p_r$, $T_r$, and $\rho_r$ are first defined as

$$p_r = p/p_c, \qquad T_r = T/T_c, \qquad \rho_r = \rho/\rho_c \tag{5-2}$$

where the subscript c refers to the property at the critical point. If two or more fluids have the same value for their reduced properties, the fluids are said to be in corresponding states. After van der Waals it is postulated that the characteristics of all fluids can be shown by an equation of state that is based on the law of corresponding states.

Equation (5–1) can be written as follows:

$$\frac{p}{\rho RT} = 1 + \frac{p_c}{\rho_c RT_c}\frac{p}{p_c}\frac{T_c}{T}B(T)\rho_c. \tag{5-3}$$

The term, $p_c/\rho_c RT_c$ is a constant which has approximately the same value for most common gases. (It is 0.292 for $O_2$ and $N_2$.) The product $B(T)\rho_c$ is approximately a function of $T_c/T$ alone for these gases. Therefore, Eq. (5–3) can be written in the form

$$\frac{p}{\rho RT} = 1 + \frac{p}{p_c}\Phi\left(\frac{T_c}{T}\right), \tag{5-4}$$

or using the definitions of Eq. (5–2)

$$Z = 1 + p_r\Phi(T_r). \tag{5-5}$$

It follows that the compressibility factor can be expressed as a function of the reduced pressure and temperature. (A more complete discussion of virial coefficients and reduced equations of state can be found in many textbooks on thermodynamics.)

The critical temperature of air is 132.5 °K, and its critical pressure is 37.2 atm. It is apparent that in gas dynamic applications the reduced pressure will rarely exceed a value of 1. For values of $T_r$ greater than 2 (corresponding approximately to normal atmospheric temperatures), the compressibility factor is small.

A number of authors (refs. 2–6) have considered the thermodynamic properties and the flow of a gas including the effect of intermolecular forces. For air at high pressures the Beattie-Bridgeman equation is considered the best empirical thermal equation of state. The disadvantage of using this equation is that it is very complex. However, in recent years, digital computors have made possible the tabulation of the thermodynamic properties and the flow characteristics of a Beattie-Bridgeman gas.

It is obvious that intermolecular forces will also affect the caloric equation of state. However, when the thermal equation of state is given, this effect is readily calculated by means of relations derived from thermodynamics. For example, Dodge (ref. 7) gives the following exact equations which form the basis for calculation of the effect of pressure on specific heats:

$$C_p - C_{p^0} = -T \int_{p^0}^{p} \frac{\partial^2 v}{\partial T^2} dp , \tag{5–6}$$

$$C_v - C_{v^0} = T \int_{v^0}^{v} \frac{\partial^2 p}{\partial T^2} dv, \tag{5–7}$$

$$C_p - C_v = T \left(\frac{\partial p}{\partial T}\right)_v \left(\frac{\partial v}{\partial T}\right)_p \tag{5–8}$$

where $v$ is the specific volume, while $p^0$ and $v^0$ refer to a state at some convenient low pressure below which the specific heat is, for all practical purposes, independent of pressure.

Randall (ref. 4) used the Beattie-Bridgeman equation to develop thermodynamic and flow process equations for gases. He included the increase in the specific heats due to the vibration of diatomic molecules by assuming the molecules to be perfect harmonic oscillators. The particular flow processes which were investigated are isentropic expansions and flow through normal shock waves. Randall's equations do not include the effect of dissociation.

Using these equations, Randall (ref. 5) calculated several of the thermodynamic properties and flow process correction factors for air. The calculations covered a range of temperatures from 70° to 2200 °R and of pressures from 0.025 to 4000 psia.

Using the equations developed by Randall, Kaufman (ref. 6) calculated the stagnation pressure and temperature, the density, the Mach number, the static pressure and temperature, and the stagnation pressure and temperature behind a normal shock. These tables cover a range of conditions in which the vibrational effects are important, but where dissociation is unimportant. It should be noted that high pressure effects are not very important for the chosen range of conditions in these tables. Therefore, the results should differ little from a similar table in which the perfect gas equation is used instead of the Beattie-Bridgeman equation of state.

## 5.3 List of Symbols

Latin symbols:

$C_p$ — specific heat at constant pressure
$C_v$ — specific heat at constant volume
$p$ — absolute pressure
$R$ — gas constant
$T$ — absolute temperature
$v$ — specific volume
$Z$ — departure coefficient

Greek symbols:

$\rho$ — density

Subscripts:

c — critical value

r — reduced value

0 — reference value

## References

1. Liepmann, H. W., and A. Roshko, *Elements of Gas Dynamics.* John Wiley, New York, 1957.
2. Crown, J. C., "Flow of a Beattie-Bridgeman Gas with Variable Specific Heat." NAVORD Report 2148, November, 1951.
3. Tao, L. N., "Gas Dynamic Behavior of Real Gases." *Journal of the Aeronautical Sciences,* Vol. 22, No. 11, November, 1955, pp. 763–764.
4. Randall, R. E., "Thermodynamic Properties of Gases: Equations Derived from the Beattie-Bridgeman Equation of State Assuming Variable Specific Heats." AEDC TR 57–10, August, 1957.
5. Randall, R. E., "Thermodynamic Properties of Air: Tables and Graphs Derived from the Beattie-Bridgeman Equation of State Assuming Variable Specific Heats." AEDC TR 57–8, August, 1957.
6. Kaufman, L. G., "Real Gas Flow Tables for Non-dissociated Air." WADC TR 59–4, January, 1959.
7. Dodge, B. F., *Chemical Engineering Thermodynamics.* McGraw-Hill, New York, 1944.

*Chapter 6*

# Dissociation and Recombination

## 6.1 Introduction

Chemical reactions, including the processes of dissociation and recombination, occur at finite rates. These reaction rates are strongly effected by temperature, and it is generally accepted that no reaction can occur unless the system of molecules and atoms possesses some minimum amount of energy, or activation energy, $E$. Until recently the kinetics of chemical reactions were of no interest to aerodynamicists, but the high temperatures associated with hypersonic flight have stimulated interest in the kinetics of dissociation and recombination, particularly for the case of both atmospheric gases and propulsion gases.

For purposes of discussion, a typical diatomic gas, $A_2$, will be considered. The dissociation and recombination processes for this molecular species may be represented as

$$A_2 + B \xrightarrow{k_D} 2A + B, \tag{6–1a}$$

$$2A + B \xrightarrow{k_R} A_2 + B. \tag{6–1b}$$

In these equations $k_D$ and $k_R$ represent the dissociation and recombination rates, respectively, and B represents an arbitrary third body necessary to either add or remove the energy necessary for the reaction to occur. For this particular case, B may be either a molecule, $A_2$, or an atom, A. The reaction rates may now be defined in terms of the concentrations, $C$, of the various species.

$$\left.\frac{dC_A}{dt}\right|_D = 2k_D\, C_{A_2}\, C_B\,, \tag{6–2a}$$

$$\left.\frac{dC_A}{dt}\right|_R = -\,2k_R\, C_A{}^2\, C_B\,. \tag{6–2b}$$

(Alternatively these reaction rates could be defined on the basis of depletion or production of the original molecular species, $A_2$.) The net rate of production of atoms may now be written as the sum of Eqs. (6–2a) and (6–2b) (note that $C_B = C_A + C_{A_2}$):

$$\frac{dC_A}{dt} = 2k_D\, C_A\; C_A + 2k_D\, C_{A_2}^2 - 2k_R\, C_A{}^3 - 2k_R\, C_A{}^2\, C_{A_2}. \tag{6–3}$$

Arrhenius first proposed an expression for chemical reaction rates. The Arrhenius equation, based largely on empirical data, is

$$k = G \exp\,(-\,E/RT). \tag{6–4}$$

Subsequent analytical expressions for chemical reaction rates generally follow this form, that is, the rate is expressed as the product of a frequency factor, $G$, and an exponential efficiency factor based on the activation energy. As will be shown later, ordinary kinetic theory is not completely adequate to predict these rates. Consequently, more sophisticated theories, such as Eyring's absolute reaction rate theory or the theory of unimolecular reactions, have been developed. However, much work, both analytical and experimental, remains before a complete description of the processes is available.

All current predictions of dissociation and recombination rates agree that they are functions of temperature only (as is their ratio, the equilibrium constant). Feldman (ref. 1) states that, in general, the dissociation rate has an exponential temperature dependence and the recombination rate has an algebraic temperature dependence.

## 6.2 Theoretical Predictions of Dissociation and Recombination Rates

### 6.2.1 Gas-Phase Reactions

#### 6.2.1.1 Simple Kinetic Theory

Kinetic theory is useful in the analysis of reaction rates if it can be postulated that the reaction occurs as a result of a collision process. Dissociation is assumed to occur when a molecule is struck by another particle possessing sufficient energy to break the molecular bond. This particle could be either another molecule or an atom. The rate at which molecules are dissociated into atoms may be considered as the product of three factors: the rate at which binary collisions occur, the fraction of these collisions in which sufficient energy is involved, and the fraction of collisions in which sufficient energy is involved that actually result in dissociation.

The number of binary collisions occuring per unit volume per unit time, $Z$, in an ideal gas may be found in most textbooks on statistical mechanics or thermodynamics. For a two-component system it is (ref. 2)

$$Z = 2\frac{C_{A_2} C_A}{\sigma_{12}} N_0 d_{A_2A} \left(\frac{2\pi RT}{\mu}\right)^{1/2}. \tag{6–5}$$

In this equation $\mu$ represents the reduced molecular weight

$$\mu = \frac{M_A M_{A_2}}{M_A + M_{A_2}}, \tag{6–6}$$

and $d_{A_2A}$ represents the effective collision diameter

$$d_{A_2A} = \tfrac{1}{2}(d_{A_2} + d_A). \tag{6–7}$$

The symmetry factor, $\sigma$, is equal to unity if the colliding particles are dissimilar and equal to 2 if the particles are similar.

There are two types of collisions of importance, atom-molecule collisions and molecule-molecule collisions. Atom-atom collisions are neglected since they cannot produce additional atoms. The total number of collisions that could result in dissociation, $Z_D$, is therefore the sum of these two types of collisions and may be written as

$$Z_D = C_{A_2}^2 (2\pi RT)^{1/2} N_0 d_{A_2}^2 \left(\frac{2}{M_{A_2}}\right)^{1/2} + 2C_A C_{A_2} (2\pi RT)^{1/2} N_0 \left(\frac{d_{A_2} + d_A}{2}\right)^2 \left(\frac{M_{A_2} + M_A}{M_{A_2} M_A}\right)^{1/2}. \tag{6-8}$$

Frost (ref. 3) notes that for a normal gas at standard conditions of temperature and pressure the number of binary collisions will be of the order of $10^{26}$ collisions cc/sec. Actually, $Z_D$ will be somewhat less at standard conditions since atom-atom collisions have not been considered. The magnitude of this difference will be directly related to the mass fraction of the dissociated species.

The postulated mechanism of dissociation includes a second factor, the fraction of the binary collisions involving an energy equal to or greater than the dissociation energy, $E_D$, of the molecule. Since an atom possesses only translational energy while a molecule possesses translational, rotational, and vibrational energies, some difficulty is encountered at this point. Convention dictates that only translational energy be considered and, based on this assumption, there are still two possible methods of calculating the relative energy of translation of the two particles. One is to consider the total relative kinetic energy between the particles and the other is to consider only that portion of the relative kinetic energy in the direction of the line of centers of the two particles at the moment of impact. The former approach gives the fraction of collisions involving sufficient energy, the energy efficiency factor, $\eta$, as

$$\eta = \exp(-E_D/RT)(E_D/RT - 1). \tag{6-9}$$

Alternatively, the later approach gives this factor as

$$\eta = \exp(-E_D/RT). \tag{6-10}$$

Fowler (ref. 2) concludes on the basis of a very limited amount of experimental data that Eq. (6–10) is the correct form for $\eta$. However, it seemed reasonable, at least for molecule-molecule collisions, that rotational and vibrational energies will contribute to the possibility of dissociation. For this reason, Eq. (6–10) is probably a conservative estimate of the energy efficiency factor.

The third factor in the rate of dissociation of molecules into atoms is the steric factor for the fraction of collisions that involve sufficient energy that actually result in dissociation. This steric factor, or probability, $P$, is a measure of the relative spatial orientation of the colliding species. It is obvious that such a term is necessary since the position of the molecule at the moment of impact must certainly influence the probability of dissociation. The steric factor will probably be different for different types of collisions. Unfortunately, kinetic theory provides no way to estimate the magnitude of the steric factor.

Combining these three factors, the net rate of dissociation of molecules may be written as

$$\left.\frac{dC_{\mathrm{A}}}{dt}\right|_{\mathrm{D}} = C_{\mathrm{A}_2}^2 (2\pi RT)^{1/2} N_0\, d_{\mathrm{A}_2}^2 \left(\frac{2}{M_{\mathrm{A}_2}}\right)^{1/2} \exp\left(-\frac{E_{\mathrm{D}}}{RT}\right) P_{\mathrm{D}_1} + \tag{6–11}$$

$$2C_{\mathrm{A}_2} C_{\mathrm{A}} (2\pi RT)^{1/2} N_0 \left(\frac{d_{\mathrm{A}_2} + d_{\mathrm{A}}}{2}\right)^2 \left(\frac{M_{\mathrm{A}_2} + M_{\mathrm{A}}}{M_{\mathrm{A}_2} M_{\mathrm{A}}}\right)^{1/2} \exp\left(-\frac{E_{\mathrm{D}}}{RT}\right) P_{\mathrm{D}_2}.$$

Comparison of Eq. (6–11) with Eq. (6–3) gives two expressions for the dissociation rate:

$$k_{\mathrm{D}} = \tfrac{1}{2}(2\pi RT)^{1/2} N_0\, d_{\mathrm{A}_2}^2 \left(\frac{2}{M_{\mathrm{A}_2}}\right)^{1/2} \exp\left(-\frac{E_{\mathrm{D}}}{RT}\right) P_{\mathrm{D}_1} \tag{6–12a}$$

$$k_{\mathrm{D}} = (2\pi RT)^{1/2} N_0 \left(\frac{d_{\mathrm{A}_2} + d_{\mathrm{A}}}{2}\right)^2 \left(\frac{M_{\mathrm{A}_2} + M_{\mathrm{A}}}{M_{\mathrm{A}_2} M_{\mathrm{A}}}\right)^{1/2} \exp\left(-\frac{E_{\mathrm{D}}}{RT}\right) P_{\mathrm{D}}. \tag{6–12b}$$

Consequently, the steric factor for molecule-molecule collisions, $P_{\mathrm{D}_1}$, must be related to the steric factor for atom-molecule collisions as

$$P_{D_2} = \frac{1}{2}\left(\frac{2d_{A_2}}{d_{A_2} + d_A}\right)^2 \left(\frac{2M_A}{M_{A_1} + M_A}\right)^{1/2} P_{D_1} \tag{6–13}$$

Since the steric factor is really a geometrical function, this relationship seems proper.

Recombination can be assumed to occur as a result of the collision of two bodies only if a third body is present to carry away the recombination energy and thus conserve momentum. A mixture of diatomic molecules and atoms can therefore recombine through one of two types of collisions. Either two atoms can collide and while they are in close proximity be struck by another atom or molecule and recombine, or else an atom and a molecule can collide and while they are in close proximity be struck by another atom and recombine. The approach used here will be that of Wood (ref. 4) in which recombination is considered to be the result of a binary collision between two colliding particles and a third particle. On this basis the analysis of the recombination process will be exactly parallel to the preceding analysis of the dissociation process. Analogous to Eq. (6–11),

$$\begin{aligned}
\left.\frac{dC_A}{dt}\right|_R = &-C_{AA}\, C_{A_2}(2\pi RT)^{1/2}\, N_0 \left(\frac{d_{AA} + d_{A_2}}{2}\right)^2 \left(\frac{M_{AA} + M_{A_2}}{M_{AA}\, M_{A_2}}\right)^{1/2} P_{R_1} \\
&- C_{AA}\, C_A(2\pi RT)^{1/2}\, N_0 \left(\frac{d_{AA} + d_A}{2}\right)^2 \left(\frac{M_{AA} + M_A}{M_{AA}\, M_A}\right)^{1/2} P_{R_2} \\
&- C_{AA_2}\, C_A(2\pi RT)^{1/2}\, N_0 \left(\frac{d_{AA_2} + d_A}{2}\right)^2 \left(\frac{M_{AA_2} + M_A}{M_{AA_2}\, M_A}\right)^{1/2} P_{R_3}
\end{aligned} \tag{6–14}$$

(The double subscripts indicate the respective particles in the process of colliding.) Equation (6–14) assumes that, since the third body in the recombination process is necessary only to absorb the energy of recombination and not furnish any energy, the energy efficiency factor is equal to unity.

The concentrations of the colliding species may be determined by assuming they are equal to the rate at which collisions occur multiplied by the mean duration of a collision. In other words,

$$C_{\mathrm{AA}} = \tau_{\mathrm{AA}} Z_{\mathrm{AA}}$$

$$C_{\mathrm{AA_2}} = \tau_{\mathrm{AA_2}} Z_{\mathrm{AA_2}}.$$

The number of collisions, $Z_{\mathrm{AA}}$ and $Z_{\mathrm{AA_2}}$, may be expressed in the form of Eq. (6–5):

$$Z_{\mathrm{AA}} = C_{\mathrm{A}}{}^2 (2\pi RT)^{1/2} N_0 \, d_{\mathrm{A}}{}^2 \left(\frac{2}{M_{\mathrm{A}}}\right)^{1/2}$$

$$Z_{\mathrm{AA_2}} = C_{\mathrm{A}} C_{\mathrm{A_2}} (2\pi RT)^{1/2} N_0 \left(\frac{d_{\mathrm{A}} + d_{\mathrm{A_2}}}{2}\right)^2 \left(\frac{M_{\mathrm{A}} + M_{\mathrm{A_2}}}{M_{\mathrm{A}} M_{\mathrm{A_2}}}\right)^{1/2}.$$

Now, the rate of depletion of atoms as a result of recombination may be written and simplified to

$$\left.\frac{dC_{\mathrm{A}}}{dt}\right|_{\mathrm{R}}$$

$$= -C_{\mathrm{A}}{}^3 (2\pi RT) N_0{}^2 \tau_{\mathrm{AA}} \, d_{\mathrm{A}}{}^2 \left(\frac{d_{\mathrm{AA}} + d_{\mathrm{A}}}{2}\right)^2 \left(\frac{2}{M_{\mathrm{A}}}\right)^{1/2} \left(\frac{M_{\mathrm{AA}} + M_{\mathrm{A}}}{M_{\mathrm{AA}} M_{\mathrm{A}}}\right)^{1/2} P_{\mathrm{R_1}}$$

$$- C_{\mathrm{A}}{}^2 C_{\mathrm{A_2}} (2\pi RT) N_0{}^2 \left[ \tau_{\mathrm{AA}} \, d_{\mathrm{A}}{}^2 \left(\frac{d_{\mathrm{AA}} + d_{\mathrm{A_2}}}{2}\right)^2 \left(\frac{2}{M_{\mathrm{A}}}\right)^{1/2} \left(\frac{M_{\mathrm{AA}} + M_{\mathrm{A_2}}}{M_{\mathrm{AA}} M_{\mathrm{A_2}}}\right)^{1/2} P_{\mathrm{R_2}} \right.$$

$$\left. + \tau_{\mathrm{AA_2}} \left(\frac{d_{\mathrm{A}} + d_{\mathrm{A_2}}}{2}\right)^2 \left(\frac{d_{\mathrm{AA_2}} + d_{\mathrm{A}}}{2}\right)^2 \left(\frac{M_{\mathrm{A}} + M_{\mathrm{A_2}}}{M_{\mathrm{A}} M_{\mathrm{A_2}}}\right)^{1/2} \left(\frac{M_{\mathrm{AA_2}} + M_{\mathrm{A}}}{M_{\mathrm{AA_2}} M_{\mathrm{A}}}\right)^{1/2} P_{\mathrm{R_3}} \right]. \tag{6–15}$$

Comparison of Eq. (6–15) with Eq. (6–3) gives two expressions for the recombination rate:

$$k_{\mathrm{R}} = \tfrac{1}{2} (2\pi RT) N_0{}^2 \tau_{\mathrm{AA}} \, d_{\mathrm{A}}{}^2 \left(\frac{d_{\mathrm{AA}} + d_{\mathrm{A}}}{2}\right)^2 \left(\frac{2}{M_{\mathrm{A}}}\right)^{1/2} \left(\frac{M_{\mathrm{AA}} + M_{\mathrm{A}}}{M_{\mathrm{AA}} M_{\mathrm{A}}}\right)^{1/2} P_{\mathrm{R_1}}, \tag{6–16a}$$

$$k_R = \tfrac{1}{2}(2\pi RT)N_0^{\,2}\Bigg[\tau_{AA}\, d_A^{\,2}\left(\frac{d_{AA}+d_{A_2}}{2}\right)^2\left(\frac{2}{M_A}\right)^{1/2}\left(\frac{M_{AA}+M_{A_2}}{M_{AA}\,M_{A_2}}\right)^{1/2} P_{R_2}$$

$$+\,\tau_{AA_2}\left(\frac{d_A+d_{A_2}}{2}\right)^2\left(\frac{d_{AA_2}+d_A}{2}\right)^2\left(\frac{M_A+M_{A_2}}{M_A\,M_{A_2}}\right)^{1/2}\left(\frac{M_{AA_2}+M_A}{M_{AA_2}\,M_A}\right)^{1/2} P_{R_3}\Bigg]. \tag{6–16b}$$

Interrelationships between $P_{R_1}$, $P_{R_2}$, and $P_{R_3}$ may be obtained by equating these two equations.

Examination of Eqs. (6–12) and (6–16) shows that simple kinetic theory will predict dissociation and recombination rates that are consistant with known criteria except for a constant, the steric factor, and, in the case of recombination, the effective collision time.

### *6.2.1.2 Theory of Unimolecular Reactions*

Another theory of chemical reaction rates is the theory of unimolecular reactions. The model, upon which this theory is based, is a polyatomic molecule in which all the oscillators are assumed to be both classical and harmonic. The molecular motion of the molecular components is described in terms of a series of internal coordinates, $\alpha_s$. For the case of a diatomic molecule, the coordinate is obviously the distance between the atoms. The reaction occurs when this length exceeds some critical value. The final result from this approach predicts a chemical reaction rate of the form (ref. 5)

$$k = \left[\sum_1^n \frac{\alpha_s^{\,2}\,\nu_s}{\alpha^2}\right]^{1/2} \exp\left(-\frac{E}{RT}\right), \tag{6–17}$$

where

$$\alpha = \sum_1^n \alpha_s^{\,2}\,. \tag{6–18}$$

In Eq. (6–17) $\nu$ is the vibrational frequency and $n$ is the number of modes of vibration. Inspection shows that the theory of unimolecular reactions predicts a rate equation similar to Arrhenius' equation. This theory is probably the most general of the existing reaction rate theories; since it is not dependent upon a Maxwell-Boltzmann energy distribution, it is difficult to apply for the actual determination of dissociation rates. Also, it will not predict recombination rates, a restriction evident from the name of the theory itself.

#### *6.2.1.3 Absolute Reaction Rate Theory*

For the simple chemical reaction

$$AB + C \rightleftarrows A + BC.$$

Eyring (ref. 6) postulates that the reaction proceeds through the formation of an activated complex, denoted by an asterisk. The total reaction equation, including this intermediate state, is therefore

$$AB + C \rightarrow (A - B - C)^* \rightarrow A + BC.$$

Jepsen and Hirschfelder (ref. 7) have examined the reaction

$$A + B + C \begin{array}{l} \rightleftarrows AB + C \\ \rightleftarrows A + BC \end{array}$$

by modifying Eyring's theory. In this case the reaction is represented as follows:

$$A + B + C \rightleftarrows (A - B - C)^* \begin{array}{l} \rightleftarrows AB + C \\ \rightleftarrows A + BC \end{array} \tag{6-19}$$

Some simplifying assumptions will be necessary in order to permit analysis of this general three-body collision process. It can be shown that a linear ensemble of atoms has the minimum energy of all possible arrangements (ref. 6). (When the formation of the activated complex involves certain electronic configurations, the state of lowest potential energy is no longer synonymous with a linear ensemble.) Hence, it is assumed that all atoms must be in such linear array before they can recombine. It is also assumed that these linear arrays are always in equilibrium with their randomly oriented neighbors. Since a chemical reaction will deplete these linear complexes, this assumption is valid only for the cases in which the rate of formation of linear complexes greatly exceeds the rate at which atoms combine. The formulation of the reaction may be broken into two steps, the formation of the linear arrays

$$\mathrm{A + B + C \rightleftarrows (A - B - C)} \qquad (6\text{–}20)$$

and the reaction of these linear arrays, through an activated complex, to form the final products

$$\mathrm{(A - B - C) \rightleftarrows (A - B - C)^{*}} \begin{cases} \rightleftarrows \mathrm{AB + C} \\ \rightleftarrows \mathrm{A + BC} \end{cases} \qquad (6\text{–}21)$$

The resulting calculation of the rates for the total reaction may be divided into two parts. First is a calculation of the equilibrium concentration of the (A — B — C) ensembles. This is dependent on quantum statistical mechanical approaches. Second is a calculation of the rate of formation of AB + C and A + BC. Included in this calculation will be a determination of the fraction of activated complexes going to each of the possible end products or back to the original species. These latter calculations involve familiar energy relationships and will consequently be considered first.

Since (A—B—C) represents a linear ensemble, it will have only two internal degrees of freedom. If the potential energy surface of this system is plotted using skewed coordinates (ref. 6), the "motion" of the system can be equivalently represented by the actual motion, in a gravitational field, of a particle moving on the potential energy surface. For the system under consideration the kinetic energy is

$$E_K = \frac{1}{2(m_A + m_B + m_C)} [m_A (m_B + m_C)\dot{r}_{AB}^2 + 2m_A m_C \dot{r}_{AB} \dot{r}_{BC} + m_C(m_A + m_B)\dot{r}_{BC}] \quad (6\text{–}22)$$

where the dot notation indicates differentiation with respect to time.

Any two non-colinear vectors on the plane can be used as a set of basic vectors for representing $r_{AB}$ and $r_{BC}$. The transformation used here is

$$r_{AB} = x - y \cot \phi \quad (6\text{–}23)$$

$$r_{BC} = \left[\frac{m_A(m_B + m_C)}{m_A(m_A + m_B)}\right]^{1/2} y \csc \phi \quad (6\text{–}24)$$

$$\phi = \cos^{-1}\left(\frac{m_C}{m_B + m_C}\right)\left[\frac{m_A(m_B + m_C)}{m_C(m_A + m_B)}\right]. \quad (6\text{–}25)$$

With this transformation, the kinetic energy, $E_k$, becomes

$$E_k = \tfrac{1}{2}\mu\dot{x}^2 + \tfrac{1}{2}\dot{y}^2, \quad (6\text{–}26)$$

where

$$\mu = \frac{m_A(m_B + m_C)}{m_A + m_B + m_C}. \quad (6\text{–}27)$$

Now the motion of a particle on the potential energy surface may be examined. Consider the surface of Fig. 6–1. For this system the particle moves to the activated state and then goes to either of the two product states. If the activated state, (A–B–C)*, has a higher potential than that of the separated reactants, the particle could also move back to the initial state.

In the general case, present knowledge is inadequate to construct the proper potential energy contours. However, surfaces have been constructed for a few specific systems (ref. 6). For these systems the dependence of the potential energy on the angle of bending has also been calculated. A complete knowledge of the four-

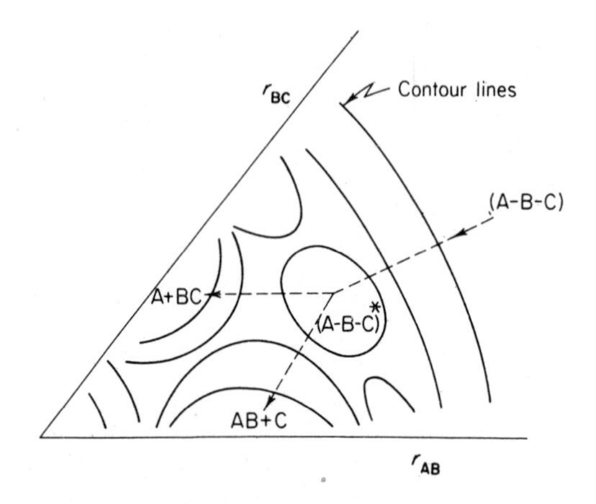


FIG. 6–1. Potential energy surface.

dimensional hypersurface is necessary for calculating the concentration of linear arrays.

Once the potential surface has been defined, the determination of all other quantities is possible, at least theoretically. Practically,

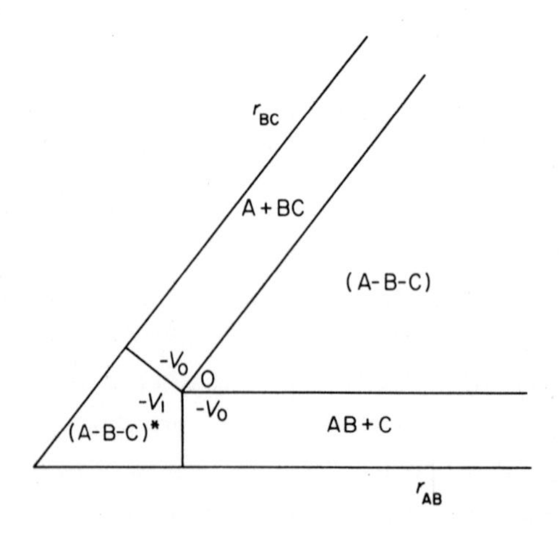


FIG. 6–2. Potential energy surface of Jepsen and Hirschfelder.

however, these calculations are extremely difficult as a result of the irregularity of the potential energy surface.

Jepsen and Hirschfelder (ref. 7) have proposed a very simple surface consisting of potential plateaus as shown in Fig. 6–2. In this highly idealized model, a particle incident upon a potential discontinuity is either reflected or refracted. The "refractive index" for a particle of energy $E$ is

$$\left[\frac{(E - V_i)}{V_i}\right]^{1/2}.$$

(Reference 7 gives a different value for this refractive index as a result of a misprint in the original paper.) The fraction of the incident particles which pass through the (A — B — C)* well and remain in either of the two product troughs, AB + C or A + BC, has been calculated for different values of $E$, $V_0$, and $V_1$. In addition to the results of these calculations, Jepsen and Hirschfelder discuss how to compute the rate of dissociation of complexes.

The concentration of the linear arrays is the next problem. Reference 7 gives a relatively complete discussion and outline of the necessary computations. The equilibrium constant for the reaction

$$\mathrm{A + B + C \rightleftarrows (A - B - C)}$$

is given, in terms of the partition functions, $Z_i$, as

$$K = \frac{\bar{Z}}{Z_A Z_B Z_C}. \tag{6–28}$$

$\bar{Z}$ is the partition function for the linear array and $Z_i$ is the partition function for the $i$th reacting species.

If it is assumed that there are only weak interactions between the molecules, the partition function may be written as the product of the partition functions representing each of the different forms of energy present: $Z = Z_{trans} Z_{vib} Z_{rot} Z_{elec}$. For a linear molecule of $j$ atoms, these individual partition functions may be written as (ref. 8)

$$Z_{\text{trans}} = V\left(\frac{2\pi mkT}{h^2}\right)^{3/2} \tag{6–29}$$

$$Z_{\text{vib}} = \prod_{i=1}^{3j-5}\left[1 - \exp\left(-\frac{h\nu_i}{kT}\right)\right]^{-1} \tag{6–30}$$

$$Z_{\text{rot}} = \frac{8\pi^2 IkT}{\sigma h^2} \tag{6–31}$$

$$Z_{\text{elec}} \cong g. \tag{6–32}$$

In the particular case of very high temperatures, Eq. (6–30) may be approximated as

$$Z_{\text{vib}} \cong \prod_{i=1}^{3j-5} \frac{kT}{h\nu_i}. \tag{6–33}$$

The electronic contribution to the partition function will equal the statistical weight of the ground electronic level when the energy of the first excited electronic level greatly exceeds $kT$. A mon-atomic molecule does not possess either rotational or vibrational energy.

According to Eyring's theory, in constructing the partition function of the complex, those terms which lead to dissociation of the complex are factored out and lumped in the calculation of the rate of dissociation of the complexes. Jepsen and Hirschfelder argue that, at high temperatures, there are two such terms, corresponding to vibrations along the axis of the linear complex. For a three-atom linear complex, there are two stable modes of vibration. If all three atoms are of the same species, then both modes will have the same frequency.

For the atomic recombination process, A = B = C, and

$$K = \frac{\bar{Z}}{Z_A{}^3},$$

or

$$K = \frac{\bar{g}\left(\frac{2\pi mkT}{h^2}\right)^{3/2}\left(\frac{8\pi^2 \bar{I}kT}{\bar{\sigma}\, h^2}\right)\left(\frac{kT}{h\bar{\nu}}\right)^2}{g_A{}^3\left(\frac{2\pi m_A\, kT}{h^2}\right)^{9/2}}. \tag{6-34}$$

This function relates the concentration of the atomic species to the concentration of the linear ensembles.

It is of interest to determine the temperature dependence of the recombination rate predicted by this absolute reaction rate theory. According to Jepsen and Hirschfelder,

$$k_R \sim \frac{\bar{Z}}{Z_A{}^3}\int_0^\infty \exp\left(-\frac{E}{kT}\right) f(E)\, dE\,, \tag{6-35a}$$

where $f(E)$ is a lumped term. Using the mean value theorem,

$$\int_0^\infty \exp\left(-\frac{E}{kT}\right) f(E)\, dE = f(\bar{E})\int_0^\infty \exp\left(-\frac{E}{kT}\right) dE\,,$$

where $\bar{E}$ is the mean value of $E$. Then

$$k_R \sim \frac{(T^{3/2})(T)(T^2)}{(T^{9/2})}\cdot\frac{1}{(1/T)}\,, \tag{6-35b}$$

or

$$k_R \sim T. \tag{6-35c}$$

Therefore, the recombination rate for diatomic molecules should be a linear function of temperature.

### *6.2.1.4 Discussion and Evaluation of Theories*

Kinetic theory is certainly not suitable to describe the reaction rates in complicated processes if quantum mechanical effects are significant. However, experimental data on the reaction rates for diatomic molecules is not sufficient at this time to evaluate the magnitude of these effects in elementary molecular structures.

Therefore, collision theory may be adequate to describe the dissociation and recombination processes for a diatomic gas and, due to its relative simplicity, certainly deserves careful examination.

The theory of unimolecular reactions, although adequate for predicting dissociation rates, is useless for predicting recombination rates as a result of the postulated reaction mechanism. This theory is definitely more general than the other two theories discussed here, since it is not dependent upon a Maxwell-Boltzmann energy distribution. However, as a result of the restrictive nature of the model and the difficulty of the calculations, this theory is the least useful in aerodynamic studies.

TABLE 6–1
RECOMBINATION RATES FOR OXYGEN

| Recombination rate ($cm^6$ $gm\text{-}mole^{-2}$ $sec^{-1}$) | Reference |
|---|---|
| $1.09 \times 10^{17}\ (T/300)^{-2}$ | 9 |
| $2.6 \times 10^{17}\ (T/300)^{-2}$ | 10 |
| $5 \times 10^{14}\ (T/300)^{-1.5}$ | 11 |
| $3 \times 10^{16}\ (T/300)^{-1.5}$ | 12 |
| $4.05 \times 10^{17}\ (T/300)^{-1.5}$ | 13 |
| $4.53 \times 10^{17}\ (T/300)^{-0.5}$ | 14 |
| $8 \times 10^{12}\ (T/300)^{0.5}$ | 15 |
| $3 \times 10^{15}\ (T/300)^{0.5}$ | 16 |
| $1.2 \times 10^{16}\ (T/300)^{2.5}$ | 17 |

The major limitation of the theory of absolute reaction rates is the lack of knowledge concerning the nature of the potential energy surfaces required. As more progress is made in this area, this theory should prove to be the most valuable for predicting the dissociation and recombination rates for most common gases.

All calculations of chemical reaction rates are dependent on numerical values for the various molecular parameters. Due to inherent experimental difficulties in measuring these values, there is considerable uncertainty in their magnitudes. Additional problems occur as a result of the fact that many of these experiments are based on momentum transfer processes, and it is questionable whether or not they are equally applicable for energy transfer processes. An ideal check on the validity of any of the reaction rate theories can be made only when sufficient experimental data are available. This current lack of data is obvious upon inspection of Table 6–1, which lists recombination rates for oxygen currently set forth in the literature.

### 6.2.2 Surface Recombination

#### *6.2.2.1 Recombination at a Catalytic Wall*

Chemical adsorption (ref. 18), most easily distinguished from physical adsorption by much higher heats of adsorption, proceeds by the formation of chemical bonds between gaseous atoms and surface atoms of the adsorbing material. It is generally believed that diatomic gases dissociate in adsorption, the evidence being the particular pressure dependence of the reaction rate. Since such a mechanism involves a cleavage of the chemical bonds between the atoms of the gas molecules, an activation energy for the adsorption reaction is implied. The assumption of activated adsorption, first made by Taylor (ref. 19), has been borne out by experiment.

A dissociated gas will be adsorbed much more readily than a molecular gas, since no activation energy is necessary. A desorbing gas is likely to desorb as molecules rather than as atoms, since the former process involves a lower activation energy. Therefore, it is likely that a surface on which adsorption proceeds by the above mechanism will act as a catalyst for the recombination reaction.

For many gases and adsorbing solids, desorption is rapid at high temperatures due to the instability of the surface compounds formed. Only a small fraction of the adsorbing surface is covered, and the recombination is a second-order reaction. At lower temperatures the surface remains almost completely covered; in this case the reaction is first order, the only concentration dependence being on the concentration of gaseous atoms.

In order to explain this behavior, a recombination mechanism has been developed by Rideal (ref. 20). This mechanism assumes a reaction between an adsorbed atom and a gas atom to form a molecule, the empty site rapidly adsorbing another gas atom. Using calculations based on such a mechanism, Laidler (ref. 21) achieved good agreement with experiment for various combinations of gases and adsorbing surfaces.

The hypothesis that the recombination reaction proceeds by the Rideal mechanism at lower temperatures and by the association of desorbing atoms at high temperatures (ref. 22) is generally borne out by calculation. Experiments by Linnett and Marsden (ref. 23) at lower temperatures indicate that the recombination reaction for oxygen or glass has a low activation energy, a fact in accord with the Rideal mechanism. At higher temperatures the surface is almost bare, and recombination by the Rideal mechanism is probably negligible.

In order to simplify calculations, it is generally assumed that the concentrations of atomic and molecular species in the gas phase are constant. Actually, this is not true, since there is dissociation and recombination in the gas phase, and the catalytic activity of the wall will lower the concentration of the atomic species and raise that of the molecular species. It is also assumed that the concentration of adsorption sites is constant over the surface of the catalyst and that the temperature of the wall is the same as the temperature of the gas. The magnitude of the error introduced by the above assumptions cannot at this time be estimated.

The reaction rate equations are derived by Eyring and his associates (ref. 6). For adsorption of a molecular gas with an immobile activated complex,

$$k = C_g \left[ n_s \frac{z}{z-\theta} (1-\theta)^2 \right] \frac{kT}{h} \left( \frac{h^2}{2\pi m kT} \right)^{3/2} \left( \frac{\sigma h^2}{8\pi^2 IkT} \right) \exp\left( -\frac{E}{kT} \right), \tag{6-36}$$

where

$$C_s = n_s \frac{z}{z-\theta} (1-\theta)^2. \tag{6–37}$$

For the adsorption of an atomic gas with an immobile activated complex,

$$k = C_g [n_s(1-\theta)] \frac{kT}{h} \left( \frac{h^2}{2\pi mkT} \right)^{3/2} \exp\left(-\frac{E}{kT}\right), \tag{6–38}$$

where

$$C_s = n_s(1-\theta). \tag{6–39}$$

For adsorption of a gas with a mobile activated complex,

$$k = C_g \frac{kT}{h} \left( \frac{h^2}{2\pi mkT} \right)^{1/2} \exp\left(-\frac{E}{kT}\right). \tag{6–40}$$

Assuming immobile activated complexes, we have for the desorption of atoms

$$k = n_s \theta \frac{kT}{h} \exp\left(-\frac{E}{kT}\right), \tag{6–41}$$

where

$$C_a = n_s \theta, \tag{6–42}$$

and for desorption of molecules

$$k = n_s \frac{(Z-1)^2}{Z(Z-\theta)} \theta^2 \frac{kT}{h} \exp\left(-\frac{E}{kT}\right). \tag{6–43}$$

For recombination by the Rideal mechanism (ref. 21),

$$k = C_g n_s \theta \frac{kT}{h} \left( \frac{h^2}{2\pi mkT} \right)^{3/2} \exp\left(-\frac{E}{kT}\right), \tag{6–44}$$

where

$$C_a = n_s \theta. \tag{6–45}$$

The expressions for $C_s$ and $C_a$ are derived by Miller (ref. 24) and are based on the fact that an adsorbing gas molecule must find two adjacent empty adsorption sites, while a desorbing molecule must involve atoms from two neighboring sites.

In order to obtain reliable results in calculations of this type, (1) one must have a firm basis for deciding whether the activated complexes are mobile or immobile at high temperatures, (2) the temperature difference between the wall must be included, and (3) the effects of heterogeneity of the catalytic surfaces should be worked into the calculation scheme.

#### *6.2.2.2 Methods for Altering Catalytic Efficiency*

The catalytic activity of a surface can be altered by either physical or chemical methods. Though the former are often very effective, they generally entail the changing of other material properties as well. For example, addition of alloying elements, addition of impurities, cold working, etc., will change the crystal structure of a metal and in many cases its catalytic properties, but will often affect the other physical characteristics of the metal. For this reason, and because chemical methods are better known, the present discussion will be concerned with chemical methods of changing catalytic activity.

The literature on catalysis shows a bewildering variety of names for substances used to lower catalytic activity. For example, Emmett (ref. 22) uses the terms "poison" and "inhibitor" to label such substances. Poisons refer to substances in the gas stream such as impurities or reaction products that have an adverse effect on catalyst performance. Inhibitors are defined as substances added during preparation of a catalyst to lower its activity. Thus poisons can be inhibitors and vice versa. To lessen confusion, this discussion will consider all such substances as inhibitors. In Griffith's terminology (ref. 25), "poisoning" and "retardation" are the two effects predominantly responsible for lowering catalytic activity. Poisoning is due to the formation of surface compounds which do not have effective catalytic properties, while retardation is due to the covering of the catalytic surface by a gas which does not take part in the reaction. Griffith's definitions are somewhat arbitrary, but are useful and will be employed below.

Selenium compounds, sulfur compounds, $O_2$, and $H_2O$ might be expected to act as poisons for the recombination of nitrogen on metals, since they act as powerful poisons toward normal nitrogen adsorption. These substances react with metals to form surface compounds which are unfavorable for adsorption. They also act as poisons for oxygen recombination, as shown directly by Greaves and Linnett (ref. 26).

Hydrogen is an excellent example of a gas whose presence would retard recombination of nitrogen on metals. Because the adsorption activation energy of hydrogen is low, it is adsorbed more readily than nitrogen and occupies much of the adsorbing surface. It should not be inferred that a gas with a lower activation energy will automatically be a retarder. Often the presence of adsorbed gases will so lower the adsorption activation energy for another gas that that gas is adsorbed more readily. No hard and fast rule may be applied.

The disadvantage of using poisons is that they may change properties of the surface coating, particularly heat transfer properties. Their strong point is their greater stability. Retarders have less of an effect on the surface but are less stable. They may tend to desorb rapidly under extreme conditions of temperature and thus lose their retarding effectiveness.

In the language of Emmett (ref. 22), "promotors" are substances added during catalyst preparation to increase catalytic activity, while "accelerators" are substances in the gas stream, such as reaction products, which increase the catalytic activity. Griffith (ref. 25) lumps both of these together as promotors.

One well-known promotor is alumina, which, when added to iron, promotes nitrogen adsorption. Since virtually every book on catalysis carries extensive discussions of promotors, no further examples will be given here. It should be noted, however, that promotors may affect material properties significantly.

## 6.3 Experimental Techniques for the Investigation of Heterogeneous Chemical Reactions

### 6.3.1 Experimental Methods

The two essentials of any recombination experiment are a means of producing gaseous atoms and a method for quantitatively detecting their presence. If the desired measurements are to be of homogeneous recombination, the walls of the system can be chemically poisoned to minimize wall reactions. Alternatively, a study of heterogeneous recombination can be made by allowing recombination on chosen solid catalysts and by keeping the system pressure low to make recombination in the gas phase negligible.

The Wood's tube method is widely used for producing gaseous atoms. Wood (ref. 27) found that a high degree of dissociation is effected by an electrical discharge through a gas and that the presence of $H_2O$ in the system prevents rapid recombination. The Wood method is not used in conjunction with high temperatures and pressures as these conditions induce excessive recombination. Recently, microwave methods (ref. 28) have been used for dissociating gases. Very high yields have been obtained using these methods.

Although electrical methods are used almost exclusively for atom production, other means of dissociation are worth comment. One of these is thermal decomposition effected by heating a wire in a diatomic gas. The gas molecules dissociate in adsorption and may come off as atoms in desorption, as first shown by Langmuir (ref. 29). The yield from such a method is ordinarily low. Various photochemical methods have also been tried without conspicuous success. Dissociation energies correspond to radiation in the intermediate and far ultraviolet; however, intense radiation sources and materials to transmit ultraviolet radiation are limited.

Because atoms must be excited for an emission spectrum to appear, use of this method presents certain difficulties. Wicke and Steiner (ref. 30) used emission of radiation by setting up electrodes in a flow system to produce weak local discharges. The spectra given off were analyzed to determine the decay of atomic species with distance from the atom-producing Wood's tube. Steiner (ref. 31) used this method to estimate the catalytic efficiency of the walls of his system.

Adsorption spectroscopy is also subject to experimental difficulties; however, it is a powerful technique for measuring concentration. An application of this method to the determination of hydrogen atom concentration was made by Preston (ref. 32).

The Wrede (ref. 33) gage, based on the different rates of diffusion of atoms and molecules, was used during the early 1930's for the absolute determination of atomic concentration. In particular, it was employed by Wicke and Steiner (ref. 30) as an absolute calibration for their spectroscopic method. The gage consists of a capillary or slit past which flows a stream of partially dissociated gas. If the opening is small enough, approximately one-tenth of a mean free path, flow into the capillary is by diffusion only. The capillary walls are coated with a metal film to ensure complete recombination inside the capillary. Since atoms diffuse faster than molecules, there is a steady-state pressure difference between the inside of the capillary and the outside gas stream. This pressure difference is used to determine the atom concentration in the gas stream. In order for the capillary orifice to be small enough to prevent mass flow into the tube, the mean free path of the gas particles outside must be fairly large. For this reason the Wrede gage has been used only with hydrogen gas, which has a larger mean free path than oxygen or nitrogen, and at low pressures. A strong criticism of the Wrede method was made by Poole (ref. 34) based on the fact that the orifices being used were not of the necessary smallness. This may account for the disuse into which the Wrede method has now fallen.

Smallwood (ref. 35) used the decrease of pressure due to recombination in a closed, static system to determine the rate of homogeneous recombination. The thermal conductivity of gases (ref. 36) has also been used to determine rates of gas-phase recombination. The luminescence of solids caused by surface recombination is discussed by Sancier, Fredericks, and Wise (ref. 37). As yet, this method has not been developed into a quantitative measure of the atomic concentration in the gas.

A number of chemical methods for measuring atomic concentration have been devised, but many of them are qualitative in nature. A quantitative method for determining the atom concentration and the homogeneous and heterogeneous rate constants has been used by Herron et al. (ref. 38) and by Harteck (ref. 39), based on work by Kistiakowsky and Volpi (ref. 40). Here, NO is added to a stream of N and $N_2$, the NO acting as a titrant. It rapidly disappears through the reaction

$$N + NO \rightarrow N_2 + O.$$

The amount of NO needed to react with the atomic nitrogen is used to determine the N concentration.

### 6.3.2 Experimental Observations

Most experimental work on atomic recombination has been carried out at moderate temperatures and at pressures in the range of $10^{-3}$ atm. This is not unexpected, since in the initial phases of an investigation, when no immediate application is apparent, convenience and ease of measurement are prime requirements. At present, data are needed on recombination at temperatures of the order 1000–2000 °K and at pressures at least one order of magnitude greater than those of previous work.

Production of atoms at high temperatures has hitherto been avoided because of the lower yield of atoms at high temperatures.

However, calculations indicate that a maximum heterogeneous recombination is achieved at moderately high temperatures and that the recombination rate decreases at extremely high temperatures, such as those desired. Though investigators have been discouraged by the low yields at temperatures in the range of 800 °K, it may well be that higher yields can be obtained at temperatures in the range of 2000 °K. Thus, from the standpoint of efficiency of production of atoms, experiments at these high temperatures are theoretically feasible.

At higher pressures the glow discharge in a Wood's tube becomes an arc. Though the atom production might not be seriously affected, deterioration of the electrodes with consequent contamination is a serious problem. This might be avoided through use of an electrodeless discharge. Use of such a method would also eliminate contamination due to vaporization of the electrodes at the high temperatures proposed.

Containment of gases at such temperatures would, of course, be a problem. The materials used for the walls of the system in past experiments are in many cases unsuitable. This is not a problem limited only to atom recombination experiments and hence will not be discussed here.

Calorimetric methods involving the heating of water due to atom recombination are clearly inapplicable for high temperature work. Perhaps the best calorimetric method would be the use of a wire, encased in some temperature-resistant material itself, coated with a highly efficient catalyst. The temperature rise of the wire due to catalytic heating of the catalyst could be observed by the use of optical methods, by the change in resistivity, or possibly by thermoelectric methods. In all cases allowance would have to be made for mechanical heat transfer from the hot gases. This might best be done by having a "poisoned" probe in the hot gases close to the catalytic probe, and measuring the temperature difference between the two probes. This difference would be due only to recombination on the catalytic probe.

Another possibility would be the use of the spectroscopic-chemical methods of Herron *et al.* (ref. 38) or of Harteck *et al.* (ref. 39). These should be applicable at high temperatures with only minor modifications.

Knowledge of the heterogeneous reactions of a mixture of atomic nitrogen and oxygen would be of great practical value. The difficulty lies in measuring the concentrations of the atomic and molecular species and of the compounds that would be formed. If catalysts were known which were highly catalytic to one of the atomic species and noncatalytic to the other, the problem would be solved. No catalysts are now known which have the desired properties. Furthermore, the chemical methods now used would distort the picture too much to give useful knowledge. Further work on this is necessary.

## 6.4 Applications

### 6.4.1 Enthalpy of Air in the Upper Atmosphere

The upper atmosphere is a great storehouse of energy which is due mainly to the absorption of solar radiation by air molecules. It is interesting to speculate whether this energy could be used in some kind of a propulsion system. To develop such a system, it is necessary to consider the nature of the energy supply, whether enough energy could be extracted to propel the system, and if this is the case, at what range of altitudes would operation be feasible. This is the recombination ramjet for low altitude earth satellites.

The energy in the atmosphere is a subject of interest for another reason. Recently Demetriades (ref. 41) proposed in-flight "refueling" for space vehicles. The space vehicle would have two propulsion systems, one for the boost phase and the other for the sustainer phase. Only propellants for the boost phase would be carried aloft, and these would be used to place the vehicle in orbit within the earth's atmosphere. Once in orbit the vehicle would scoop in and store whatever species are present at that altitude. When a sufficient amount had been stored, the vehicle could resume its mission (e.g., flight to Mars). The material scooped from the atmosphere would be used as propellant for a plasma or ion rocket engine during the sustainer phase. In order to store the propulsive fluid efficiently, it would be necessary to compress or condense it. The difference in energy between the incoming species and the condensed species must be removed and dumped to a heat sink or used to propel the vehicle. In any event, it is necessary to know the enthalpy of the atmosphere.

Most of the energy in the upper atmosphere is stored as the heats of formation of the dissociated and ionized species. For example, by the process of photodissociation, the oxygen molecule above heights of 100 km has been reduced largely to monatomic oxygen atoms by the reaction

$$O_2 + h\nu \rightarrow O + O \tag{6–46}$$

where $h$ is Planck's constant and $\nu$ is the frequency of the absorbed solar radiation.

If it were possible to develop a system within which the oxygen atoms in the upper atmosphere could recombine, then energy could be utilized according to the formula

$$O + O + B \rightarrow O_2 + \text{energy} + B. \tag{6–47}$$

Energy can be obtained from similar reactions involving nitrogen and ionized particles. At high altitudes increased amounts of energy are stored as sensible enthalpy as a result of the increased temperature.

The expression for the total enthalpy is given by

$$h_T = \sum_i X_i\left(\Delta H^0_{f_i} + \int_{T_0}^{T} C_{pi}\, dT\right). \tag{6–48}$$

where $X$ is the mole fraction, $\Delta H_f{}^0$ is the heat of formation and $C_p$ is the specific heat at constant pressure.

The sensible enthalpy, $h_s$, is given by the expression

$$h_s = \sum_i X_i \int_{T_0}^{T} C_{pi} \, dT , \tag{6–49}$$

$h_O$, the enthalpy contribution due to the heat of formation of the oxygen atoms, is given by

$$h_O = X_O \, \Delta H^0_{f_O} \tag{6–50}$$

where subscript O indicates oxygen. Similarly, $h_N$ and $h_I$, the enthalpy contributions due to the heats of formation of the nitrogen atoms and the "average ion," respectively, are given by

$$h_N = X_N \, \Delta H^0_{f_N} \tag{6–51}$$

$$h_I = X_I \, \Delta H^0_{f_I}. \tag{6–52}$$

It should be noted that $h_T = h_S + h_O + h_N + h_I$.

Figure 6–3 shows the volumetric enthalpy of air as a function of altitude. This parameter is of prime importance in determining the feasibility of a propulsion system. Consider, for example, the simple case of an engine with a normal shock diffuser of area $A$. The engine travels at a velocity $V$. The volume flow rate of air entering the engine is $AV$ and the rate of enthalpy intake would be $h_v \, AV$. The equation used in calculating the volumetric enthalpy is

$$h_v = h_T \frac{\rho}{M} \tag{6–53}$$

where $\rho$ is the density and $M$ is the molecular weight. Values for the quantities used in the above calculations may be found in refs. 42 and 43. Miller gives tables of atmospheric properties as functions of altitude which are based on the 1956 ARDC model atmosphere. It does not include any information on ion densities; however, Krassovsky gives the free electron density as a function of altitude (ref. 44). For the values presented here, it has been assumed that there are no doubly ionized particles present. It was also assumed that the ratio of nitrogen ions to oxygen ions was constant at a value of 0.05 (ref. 45).

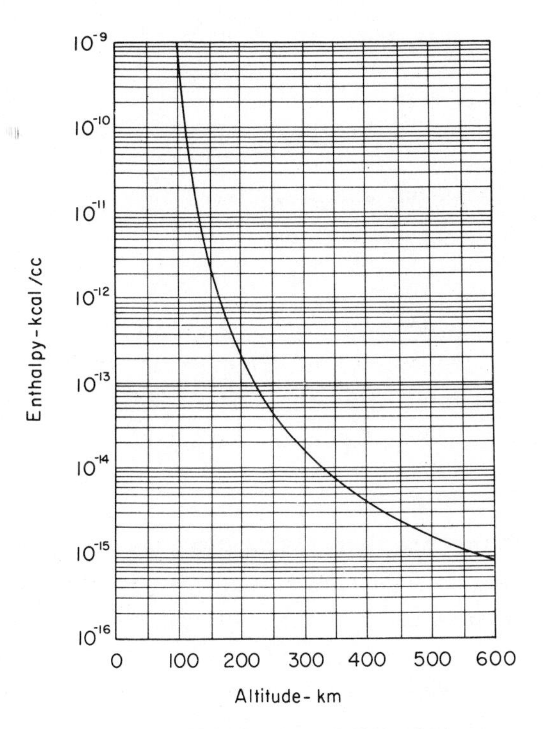


FIG. 6–3. Volumetric enthalpy of air.

## 6.4.2 Recombination Ramjet

### *6.4.2.1 Theoretical Considerations*

An illustration of the use of the energy stored in the upper atmosphere is the ramjet design. Thrust is provided by imparting a higher velocity to the exit gases relative to the engine than that of the incoming gases. The thrust is then given by the expression (assuming ideal expansion)

$$F = \dot{m}(V_e - V_i) \tag{6–54}$$

where $\dot{m}$ is the mass flow rate through the engine and the subscripts i and e refer to inlet and exit gases, respectively.

Assuming adiabatic, nonviscous flow through the engine, the first law of thermodynamics may be written as

$$\tfrac{1}{2}M_i(V_e^2 - V_i^2) = h_{T_i} - \frac{M_i}{M_e} h_{T_e}. \qquad (6\text{–}55)$$

The mass flow rate is

$$\dot{m} = A \rho_i V_i. \qquad (6\text{–}56)$$

Combining Eqs. (6–55) and (6–56),

$$F = A \rho_i V_i \left\{ \left( \frac{2(h_{T_i} - h_{T_e} M_i/M_e)}{M_i} + V_i^2 \right)^{1/2} - V_i \right\}. \qquad (6\text{–}57)$$

All quantities except $h_{T_e}$ and $M_e$ are known. The values of these two parameters depend on the characteristics of the engine. These values are determined by the effectiveness of the engine in extracting energy from the incoming gases.

### 6.4.2.2 *Test Facilities*

In spite of the fact that several investigators (see for example refs. 13 and 46) have concluded that the recombination engine is marginal as a propulsion device, interest still runs high. It is intriguing to consider the possibility of using energy stored in the upper atmosphere. Since there are so many advantages to this scheme, the feasibility should be thoroughly explored. If and when the recombination engine reaches the hardware stage, test facilities must be available. In these facilities the earth's atmosphere at 90 to 150 km must be duplicated for a vehicle flying at satellite velocity. The conditions to be duplicated are approximately Mach number 15, mole fractions of atomic oxygen up to about 0.3, temperature 530 °R, pressure ratio $p/p_0$ of $4.8 \times 10^{-6}$, and density ratio $\rho/\rho_0$ of $4.4 \times 10^{-6}$. Subscript zero means sea level value in the above ratios. The air is not at thermodynamic equilibrium; therefore, the test facility must duplicate a specified nonequilibrium condition of the gas.

The problem can be broken down into two parts: the nozzle and the high temperature gas supply. At present there appears to be two possibilities for a high energy gas supply, these being the shock tube and the hyperthermal plasma tunnel. Although magnetically driven shock waves will be considered, first glance seems to indicate that test times are too short. The length of test time for a pressure driven shock tube may be too short also. If so, then the choice appears to be narrowed to the plasma heated tunnel.

In the nozzle the flow may involve reacting gases until a certain pressure-temperature combination is attained. After that the flow will be at frozen or unchanging compositions. One aspect of the research is to apply Bray's work (ref. 47), or an extension to Bray's work, to the nozzle flow. Bray considers the location in the nozzle beyond which the gas composition is frozen. The point at which the reactions cease and the composition remains fixed must be known, since the conditions in the test section are fixed and it is necessary to work upstream to find the stagnation or nozzle entrance conditions.

The stagnation conditions then define the requirements on the hot gas generator. The source must provide hot gases at roughly 7000 °K and at pressures up to 10 atm. For the conditions specified the stagnation pressure is the order of 1 atm; it is conceivable that a hot gas source operating up to 10 atm will be required for other test conditions. At these conditions the equilibrium composition of air is roughly that shown in the tabulation.

| Species | $O_2$ | O | $N_2$ | N | NO | A |
|---|---|---|---|---|---|---|
| Mole fraction | 0.011 | 0.260 | 0.602 | 0.046 | 0.073 | — |

To use a plasma generator, the arc must operate up to pressure levels of 10 atm.

One important question concerns the size of the test facility. Since the primary interest is in the recombination of dissociated oxygen, chemical processes in the engine must be scaled. For surface reactions the scaling laws would be quite different from those for gas phase reactions. It is necessary to scale the fluid flow also.

Short-run-time test facilities have been satisfactory for some studies of aerodynamic heating. The recombination engine differs in several ways. It is necessary to fill the recombination "combustion" chamber and establish steady state flow *within* the engine. This may require test times of sufficient length to rule out shock tubes as a possible hot gas source.

If test facilities are used which involve shock tubes, the instrumentation to determine what is going on inside the model recombination engine will be complex. A steady flow test tunnel would ease the instrumentation problem. Among the available techniques are the use of sampling probes, and absorption and emission spectroscopy.

Further research may show that it is not possible to duplicate all variables. A choice must be made as to the conditions which will be simulated. From a chemical reaction point of view, both pressure and temperature are important because the reaction rate is a function of both $T$ and $p$.

## 6.5 List of Symbols

Latin symbols:

$A$ — area
$A_2$ — typical diatomic molecule
B — arbitrary third body
$C$ — concentration
$C_p$ — specific heat at constant pressure
$d$ — molecular diameter

$E$ — energy
$F$ — thrust
$g$ — number of electronic levels (statistical weight)
$G$ — frequency factor
$h$ — enthalpy; Planck's constant
$I$ — molecular moment of inertia
$j$ — number of atoms in linear molecule
$k$ — Boltzman's constant; chemical reaction rate
$m$ — mass
$M$ — molecular weight
$n$ — number of modes of molecular vibration
$n_s$ — initial concentration of adsorption sites
$N_0$ — Avagadro's number
$p$ — absolute pressure
$P$ — steric factor
$r$ — distance between atoms
$R$ — gas constant
$t$ — time
$T$ — absolute temperature
$V$ — velocity
$X$ — mole fraction
$Z$ — fraction of near neighbors of any adsorption site; partition function
$z$ — number of binary collisions per unit volume per unit time

Greek symbols:

$\alpha_s$ — molecular coordinate (internal)
$\Delta H_f$ — heat of formation
$\eta$ — energy efficiency factor
$\theta$ — fraction of adsorption sites occupied
$\mu$ — reduced molecular weight
$\nu$ — vibrational frequency
$\rho$ — density
$\sigma$ — symmetry factor
$\tau$ — effective collision time

Subscripts:

a — adsorbed species
A — atomic species
$A_z$ — molecular species
D — dissociation
e — exit condition
elec — electronic
g — gaseous species
i — inlet condition
I — average ion
k — kinetic
N — nitrogen
O — oxygen
R — recombination
rot — rotational
s — available adsorption sites
T — total
trans — translational
v — volumetric
vib — vibrational

Superscripts:

* — activated
— — linear array
. — derivative with respect to time

## References

1. Feldman, S., "The Chemical Kinetics of Air at High Temperatures: A Problem in Hypersonic Aerodynamics." AVCO Research Report 4, February, 1957.
2. Fowler, R. H., and E. A. Guggenheim, *Statistical Thermodynamics*. Macmillan, New York, 1939.
3. Frost, A. A., and R. G. Pearson, *Kinetics and Mechanism*. John Wiley, New York, 1953.
4. Wood, G. P., "Calculations of the Rate of Thermal Dissociation of Air behind Normal Shock Waves at Mach Numbers of 10, 12, and 14." NACA TN 3634, April, 1956.

5. Trotman-Dickenson, A. F., *Gas Kinetics*. Academic Press, New York, 1955.
6. Glasstone, S., K. J. Laidler, and H. Eyring, *The Theory of Rate Processes*. McGraw-Hill, New York, 1941.
7. Jepsen, D. W., and J. O. Hirschfelder, "Idealized Theory of the Recombination of Atoms by Three-Body Collisions." *The Journal of Chemical Physics*, Vol. 30, No. 4, April, 1959, pp. 1032–1044.
8. Dole, M., *Introduction to Statistical Thermodynamics*. Prentice-Hall, Englewood Cliffs, New Jersey, 1954.
9. Matthews, D. L., "Interferometric Measurement in the Shock Tube of the Dissociation Rate of Oxygen." *The Physics of Fluids*, Vol. 2, No. 2, March-April, 1959.
10. Byron, S. R., "Interferometric Measurement of the Rate of Dissociation of Oxygen Heated by Strong Shock Waves." Ph. D. Dissertation, Cornell Univ., Ithaca, New York, 1958.
11. Fay, J. A., and F. R. Riddel, "Theory of Stagnation Point Heat Transfer in Dissociated Air." AVCO Research Report 18, April, 1957.
12. Demetriades, S., and C. Kretschmer, "The Use of Planetary Atmospheres for Propulsion." Presented at the Fourth Annual Meeting of the American Astronautical Society, January 31, 1958.
13. Charwat, A. F., "Photochemistry of the Upper Atmosphere as a Source of Propulsive Power." *ARS Journal*, Vol. 29, No. 2, February, 1959, pp. 108–114.
14. Heims, S. P., "Effect of Oxygen Recombination on One-Dimensional Flow at High Mach Numbers." NACA TN 4144, January, 1958.
15. Hirschfelder, J. O., C. F. Curtiss, and D. E. Campbell, "The Theory of Flame Propagation." *The Journal of Physical Chemistry*, Vol. 57, No. 3, March, 1953, pp. 403–414.
16. Logan, J. G., "Relaxation Effects in Hypersonic Flow." IAS Preprint No. 728, January, 1957.
17. Hirschfelder, J. O., "Heat Transfer in Chemically Reacting Mixtures, I." *The Journal of Chemical Physics*, Vol. 26, No. 2, February, 1957, pp. 274–281.
18. Trapnell, B. M. W., *Chemisorption*. Academic Press, New York, 1955.
19. Taylor, H. S., "The Activation Energy of Adsorption Processes." *The Journal of the American Chemical Society*, Vol. 53, No. 2, February, 1931, pp. 578–597.
20. Rideal, E. K., "A Note on a Simple Molecular Mechanism for Heterogeneous Catalytic Reactions." *Proceedings of the Cambridge Philosophical Society*, Vol. 35, Part 2, January, 1939, pp. 130–132.
21. Shuler, K. E. and K. J. Laidler, "The Kinetics of Heterogeneous Atom and Radical Reactions, I. The Recombination of Hydrogen Atoms on Surfaces." *The Journal of Chemical Physics*, Vol. 17, No. 12, December, 1949, pp. 1212–1217.
22. Emmett, R. H. (editor), *Catalysis*. Reinhold, New York, 1954.
23. Linnett, J. W., and D. G. H. Marsden, "The Kinetics of the Recombination of Oxygen Atoms at a Glass Surface." *Proceedings of the Royal Society*, Series A, Vol. 234, No. 1199, March 6, 1956, pp. 489–504.

24. Miller, A. R., "The Heat of Adsorption of Diatomic Molecules." *Proceedings of the Cambridge Philosophical Society*, Vol. **43**, Part 2, April, 1947, pp. 232–239.
25. Griffith, R. H., and J. D. F. Marsh, *Contact Catalysis*, 2nd edition. Oxford Univ. Press, London and New York, 1957.
26. Greaves, J. C., and J. W. Linnett, "The Recombination of Oxygen Atoms at Surfaces." *Transactions of the Faraday Society*, Vol. **54**, Part 9, September, 1958, pp. 1323–1330.
27. Wood, R. W., "Spontaneous Incandescence of Substances in Atomic Hydrogen Gas." *Proceedings of the Royal Society*, Series A, Vol. **102**, No. 714, October 2, 1922, pp. 1–8.
28. Shaw, T. M., "Dissociation of Hydrogen in a Microwave Discharge." *The Journal of Chemical Physics*, Vol. **30**, No. 5, May, 1959, pp. 1366–1367.
29. Langmuir, I., "Chemical Reactions at Very Low Pressures. II. The Chemical Clean-up of Nitrogen in a Tungsten Lamp." *The Journal of the American Chemical Society*, Vol. **35**, No. 8, August, 1913, pp. 931–945.
30. Steiner, W., and F. W. Wicke, "Kinetic der Vereinigung der H-Atome im Dreierstotz." *Zeitschrift für physikalische Chemie*, Bodenstein Festband, 1931, p. 817.
31. Steiner, W., "The Loss of Hydrogen Atoms on Water-Poisoned Glass Surfaces." *Transactions of the Faraday Society*, Vol. **31**, Part 8, August, 1935, pp. 962–966.
32. Preston, W. M., "A Spectrographic Method for the Measurement of the Rate of Recombination of Atomic Hydrogen." *The Physical Review*, Vol. **57**, No. 11, June 1, 1940, p. 1074.
33. Wrede, E., "Konzentration des automaren Wasserstoffs in der Glimmentladung." *Zeitschrift für Instrumentenkunde*, Vol. **48**, No. 5, May, 1928, pp. 201–202.
34. Poole, H. G. "Atomic Hydrogen." *Proceedings of the Royal Society*, Series A, Vol. **163**, No. 914, December 7, 1937, pp. 404–454.
35. Smallwood, H. M. "The Rate of Recombination of Atomic Hydrogen." *Journal of the American Chemical Society*, Vol. **56**, No. 7, July, 1934, pp. 1542–1549.
36. Senftleben, H., and E. Germer, "Nachweis einer durch direkte Bestrahlung bewirkten Dissoziation der Halogenmoleküle." *Annalen der Physik*, Vol. **2**, No. 7, September 4, 1929, pp. 847–864.
37. Sancier, K. M., W. J. Fredericks, and H. Wise, "Luminescence of Solids Produced by Surface Recombination of Atoms." *The Journal of Chemical Physics*, Vol. **30**, No. 5, May, 1959, pp. 1355–1356.
38. Herron, J. T., J. L. Franklin, P. Bradt, and V. H. Dibeler, "Kinetics of Nitrogen Atom Recombination." *The Journal of Chemical Physics*, Vol. **30**, No. 4, April, 1959, pp. 879–885.
39. Harteck, P., R. R. Reeves, and G. Mannella, "Rate of Recombination of Nitrogen Atoms." *The Journal of Chemical Physics*, Vol. **29**, No. 3, September, 1958, pp. 608–610.
40. Kistiakowsky, G. B., and G. G. Volpi, "Reactions of Nitrogen Atoms. II. $H_2$, CO, $NH_3$, NO, and $NO_2$." *The Journal of Chemical Physics*, Vol. **28**, No. 4, April, 1958, pp. 665–668.

41. Demetriades, S. T., "Comments on 'Photochemistry of the Upper Atmosphere as a Source of Propulsive Power.' " *ARS Journal,* Vol. 29, No. 5, May, 1959, p. 375.
42. Miller, L. E., "Molecular Weight of Air at High Altitudes." *Journal of Geophysical Research,* Vol. 62, No. 3, September, 1957, pp. 351–365.
43. Penner, S. S., *Chemistry Problems in Jet Propulsion.* Pergamon Press, New York, 1957.
44. Krassovsky, V. I., "Exploration of the Upper Atmosphere with the Help of the Third Soviet Sputnik." *Proceedings of the IRE,* Vol. 47, No. 2, February, 1959, pp. 289–296.
45. Friedman, H., "Rocket Observations of the Ionosphere." *Proceedings of the IRE,* Vol. 47, No. 2, February, 1959, pp. 272–279.
46. Baldwin, L. V., and P. L. Blackshear, "Preliminary Survey of Propulsion Using Chemical Energy Stored in the Upper Atmosphere." NACA TN 4267, May, 1958.
47. Bray, K. N. C., "Departure from Dissociation Equilibrium in a Hypersonic Nozzle." ARC 19, 983, March 17, 1958.

## Chapter 7

# Ionization and Neutralization

## 7.1 Introduction

Emphasis in this chapter has been placed on the fundamental processes of ionization and recombination, since these processes must be understood before more complicated phenomena can be considered. The individual discussion of each phenomenon is necessarily brief, since an effort was made to survey the complete field. Further details may be obtained by consulting the references listed at the end of the chapter. Although some of the processes described may not apply explicitly to problems of an aerodynamic nature, they are included for the sake of completeness.

Some experimental data are used to illustrate particular points. However, the reader is cautioned about imprudent applications of experimental information. Since the number of possible phenomena is great, one must be certain that the experimental data and the process under consideration are compatible. Also, since impurities in the gas have drastic effects on the data, much of the experimental results are of questionable accuracy.

## 7.2 Ionization Processes

Ionization is the process whereby one or more electrons are removed from an atom. Since this reaction absorbs energy, energy must be supplied to the atom. The energy required to remove a given electron from its atomic orbit and place it at rest at an infinite distance is defined as the ionization potential, $E_{\mathrm{I}}$

(or more correctly, ionization energy), of the atom. It is usually expressed in electron volts. The first ionization potential is the energy required to remove the first electron from the atom, the second ionization potential is the energy required to remove the second electron, etc. The values of the first ionization potential range from 3.88 ev for cesium to 24.46 ev for helium. The values of the second ionization potential range from 9.95 ev for barium to 54.14 ev for helium. The succeeding ionization potentials are correspondingly higher. Consequently, at low temperatures doubly ionized atoms are less probable than singly ionized atoms. Tables of the ionization potentials of the elements can be found in handbooks.

Ionization processes can occur either singly or in combination. These processes can be classified as follows:

(1) Ionization by electron collisions
(2) Ionization by positive ion or neutral atom collisions
(3) Ionization by radiation
(4) Ionization by collisions of the second kind
(5) Cumulative ionization
(6) Thermal ionization
(7) Surface ionization

In addition, electron attachment can occur. Although this is not an ionization process, the end result is similar, i.e., the transformation of a neutral particle into a charged particle.

### 7.2.1 Ionization by Electron Collisions

The ionization energy may be given to an atom by collision with an electron. This is perhaps the most common ionization process. The probability of an electron ionizing an atom depends on the cross section for ionization which is a function of electron energy. If an atom is to be ionized in a single collision, the minimum relative energy of the electron and atom must be the ionization energy of the atom. Less energetic systems having less than the ionization energy can raise an atom to an excited state. This atom may then be ionized in a subsequent collision with another electron.

It is customary to plot the number, $S$, of ion pairs produced by an electron in traveling 1 cm through a gas at a pressure of 1 mm Hg as a function of the initial energy of the electron (see Fig. 7–1). This quantity is often called the probability of ionization,

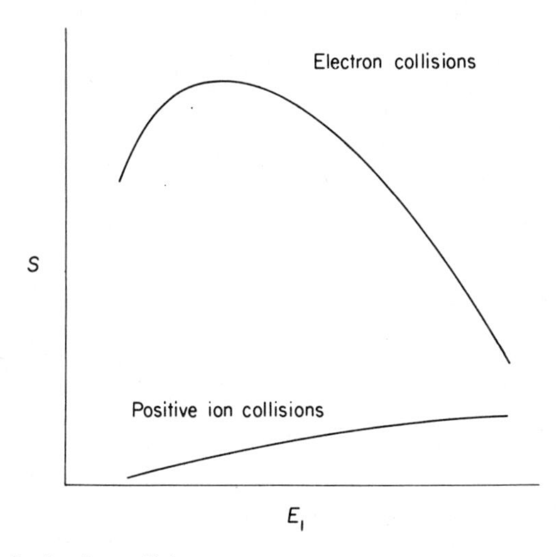


FIG. 7–1. Ionization efficiency versus electron energy (collision process).

but since the dimensions of $S$ are not those of probability, the terms differential ionization coefficient, ionization efficiency, and ionizing ability are also used.

It is found that for single ionization $S$ reaches a maximum for initial electron energies about 5 to 10 times the ionization energy of the atom, and continues to decrease for higher electron energies. The value of $S$ for double, triple, etc., ionization of the atom reaches a peak at correspondingly higher values of electron energy (ref. 1).

An electron can lose energy in nonionizing collisions before it ionizes an atom. Therefore, it is sometimes useful to define an effective ionization potential as the ratio of the initial electron energy to the number of ion pairs formed per electron. The effective ionization potential will obviously be greater than the first ionization potential of the atom.

If the colliding electron has more than the minimum amount of energy required to ionize the atom, the excess energy may be retained by the electron, transferred to the electron stripped from the atom, or used to excite or further ionize the atom. A combination of these possibilities may also result. Generally, the excess energy transformed into kinetic energy of the ion will be insignificant. In many polyatomic molecules, the internal energy of the resulting positive ion is greater than that of its dissociated products (ref. 2). Therefore, dissociation of the ionized molecule may result. Usually, this secondary process requires an external stimulus, such as a collision, even though it is possible without external aid. For example, in hydrogen, the first product is $H_2^+$. Upon collision with other molecules, this ion may dissociate into $H^+ + H$ or associate to $H_3^+ + H$.

### 7.2.2 Ionization by Positive Ion or Neutral Atom Collisions

Compared to electrons, positive ions are generally inefficient in producing ionization by collisions. Some very energetic positive ions, such as alpha particles, however, are effective ionizing agents. Because its mass is of the order of the mass of the atoms or molecules of the gas, a positive ion will not attain an energy much larger than that of the neutral gas particles unless it is in a strong electric field or a very dilute gas. Consequently, except for very high temperature gases, most of the positive ions will have an energy less than the ionization energy of the gas molecules, and ionization by collision will be improbable.

It is found that positive ions begin to be effective ionizers when their velocities are as great as those of electrons having an energy equal to the ionizing energy of the gas. Since it is unlikely that the positive ions will attain such a velocity, ionization by positive ion collision is probably less important than any secondary effects which may occur. These secondary effects can be secondary emission of electrons from walls and electrodes, or inelastic collisions with gas atoms in which the radiation emitted causes photoelectric emission in the gas or at surfaces (ref. 3).

When argon is bombarded with positive ions of the alkali metals (see Fig. 7–1), the ionization efficiency of the ions is much less than that of electrons for energies up to 1000 ev (ref. 4).

Ionization by collisions with neutral atoms is considered an important process, although comparatively little is known about it. Investigations have shown that for atoms of the same element, ionization is possible if the relative energy is at least twice the ionization energy. Because of the high energies required, this process can only occur in very high temperature gases, such as those encountered in aerodynamic studies.

This process is about as efficient as ionization by positive ions (ref. 5). The process is somewhat less efficient, however, if the ionizing atoms are of a different element than those of the gas.

### 7.2.3 Ionization by Radiation

A photon may excite or ionize an atom if its energy is greater than the ionization energy, $E_I$, of the atom. The energy of radiation is given by $h\nu$ where $h$ is Plank's constant and $\nu$ is the frequency of the radiation. The ionization efficiency of photons is a maximum for radiation energies just slightly above the ionization energy of the atom, and decreases very rapidly for higher energies. The ionization efficiency also depends on the radiation density.

Very high frequency radiation, instead of being absorbed by the outermost electrons of the atom, acts preferentially on an electron in an inner shell (ref. 5). Radiation will, of course, be emitted when this vacancy in the inner shell is filled by an electron from a higher level. Radiation will also be emitted from any atom, which has been raised to an excited state, when it returns to its normal state. Although it is very common, ionization by photons is generally of less importance than ionization by electrons.

The lowest frequency photon that will ionize an atom can easily be calculated. Substituting the lowest ionization potential, 3.88 ev for cesium into $\nu = E_I/h$, $\nu$ is found to be $9.33 \times 10^{14}$ cps which corresponds to a wavelength of 3210 A. The required radiation is well into the ultraviolet region of the spectrum and therefore visible light will not ionize an atom although it may excite it. However, emission due to visible light will take place at certain surfaces, since here the required energy (work function) is much smaller.

The absorption of energy from a beam of radiation by a gas and its subsequent reradiation in any direction causes the intensity of the beam to decrease with the distance traversed. The intensity $I$ of radiation as a function of distance $x$ is given by (ref. 1)

$$I = I_0 \exp(-\mu x)$$

where $I_0$ is the intensity of the entering beam of photons, and $\mu$ is known as the absorption coefficient of the gas. This coefficient is a characteristic of the gas and is a function of the frequency of the incident radiation.

### 7.2.4 Ionization by Collisions of the Second Kind

An atom can be ionized by the transfer of energy from an excited or ionized atom. This process is possible if the available energy is equal to the ionization energy of the atom. The probability of the occurrence of this process is a maximum if the available energy is just sufficient, and decreases with an excess of energy (ref. 2).

### 7.2.5 Cumulative Ionization

This process refers to ionization in two or more stages. An atom excited by one method may have its ionization completed by one or more other processes. This process may include collisions of the second kind discussed previously. Collisions of the second kind usually occur when gases are mixed, since the energy of radiation emitted when an excited atom of one species returns to its normal state may be greater than the excitation energy of another species. In cases where the energy is not sufficient, the process may be effected by the absorption of part of the kinetic energy of the excited atom.

Metastable states provide the most important source of excitation by excited atoms. Metastable states are those from which spontaneous escape by radiation is inhibited according to the selection principles of quantum theory. Therefore, the lifetime of this state is much longer than a normal excited state, lasting as long as $10^{-1}$ sec compared with the average life of a normal excited state of the order of $10^{-8}$ sec. In a lifetime of this length, the atom can undergo many collisions, and the probability of energy transfer is consequently high. Metastable atoms can have a considerable effect on the conductivity of a gas if it contains atoms of a different element with an ionization energy less than the energy corresponding to the metastable excitation level of the gas atoms (ref. 6).

The metastable state can be destroyed by one of the following processes: (a) The atom may be excited to a higher energy level by cumulative excitation by any one of the previously mentioned processes. (b) The atom may undergo a collision of the second kind with an atom of the same element so that its excitation energy plus its kinetic energy can excite the atom to a higher level. (c) In mixed gases, the energy of the atoms in the metastable state of one gas may go into the ionization of the atoms of another gas if the available energy is high enough. (d) Metastable atoms may collide with solid surfaces and cause secondary emission of electrons.

### 7.2.6 Thermal Ionization

This process occurs when the average kinetic energy of the molecules is high enough so that the energy transferred in a collision between two neutral molecules is sufficient to ionize one of them. The name, thermal ionization, probably results from the fact that this process occurs only at very high temperatures, but is somewhat misleading since the process is still one of ionization by collision. In the past, this process was encountered in electric arcs and in high temperature flames. In recent years, it has been encountered behind strong shock waves, in boundary layers, in hypersonic flow fields, and in processes connected with thermonuclear energy research.

Ionization by collision of neutral particles does not occur alone, however. When the number of electrons in the gas becomes appreciable, ionization by electrons may become predominant, since electrons are much more efficient than neutral atoms or ions as ionizing agents. In addition, radiation emitted from the walls and from the gas itself may be significant.

Under equilibrium conditions, the number of ions lost by recombination will equal the number formed by the processes of thermal ionization. This very complicated situation can be simplified for the purposes of analysis by assuming that the ionization process is a completely reversible reaction defined by the equation:

$$\mathrm{A} + E_{\mathrm{I}} \rightleftarrows \mathrm{A}^{+} + e \tag{7-1}$$

where A represents the neutral atom, $\mathrm{A}^{+}$ the single ionized atom, $e$ the electron, and $E_{\mathrm{I}}$ the ionization energy. This equation is valid for a monatomic gas or for the atoms of a diatomic gas which is completely dissociated. At equilibrium, the particles will possess a Maxwell-Boltzmann distribution, and will have a mean kinetic energy corresponding to the temperature $T$. The equilibrium concentrations of the particles are given by the Saha equation, Eq. (3–71) and discussed in Chapter 3. Figure 7–2 shows a representative calculation for the degree of ionization $\phi$ as a function

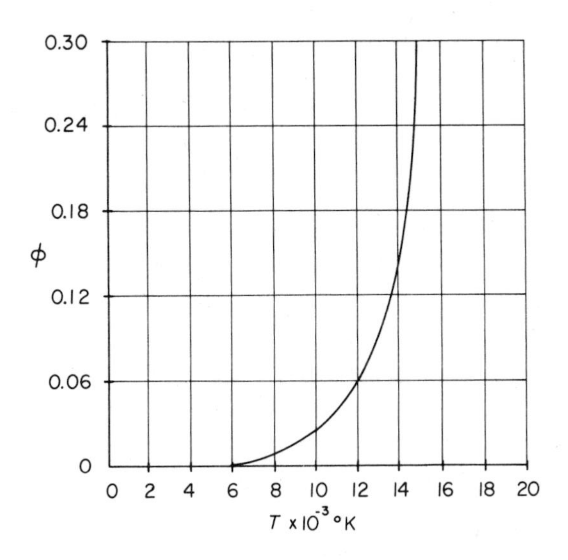


FIG. 7–2. Fraction of molecules thermally ionized — air at 1 atm pressure.

of temperature for air. It can be seen that there is negligible ionization at temperatures less than approximately 6000 °K. Appropriate equations which are required when multiple ionization must be considered are also given in Chapter 3.

It should be noted that ionized gases are often not in thermal equilibrium due to finite relaxation times. A typical example is the region immediately behind a strong shock wave. For these cases, the various ionization and recombination mechanisms must be considered, and the appropriate reaction rate equations must be used.

### 7.2.7 Surface Ionization

When a neutral atom strikes a hot surface, it may lose its outermost electron to the surface, and bounce off as an ion. This process is known as surface ionization and is a type of thermal ionization. The ionization efficiency (i.e., the ratio of the number of ions coming from the surface to the number of incident atoms) can be predicted, at least qualitatively, from a modified form of the Saha equation known as the Saha-Langmuir equation (see ref. 7). According to this equation, the efficiency is proportional to $\exp{(\phi' - E_\text{I})/kT}$ where $\phi'$ is the thermionic work function of the surface. The equation does not apply for low temperatures, however, since in the actual case, a threshold temperature must be reached before any appreciable ionization takes place. This is caused by a deposit of neutral atoms on the surface which lowers the work function of the surface to that of the element to be ionized. At the treshold temperature, the atoms begin to boil off the surface, and ionization then begins to take place.

The work function of the surface must be greater than the ionization potential of the atom to be ionized for appreciable ionization to take place. Since the highest work functions are of the order of 6 ev, only the alkali metals can be ionized by this process with high efficiency.

Reference 7 presents a summary of the theory plus some experimentally determined results for the ionization efficiency of alkali metals on tungsten and platinum surfaces.

The usefulness of surface ionization lies in the fact that a very simple ion source can be made for the ionization of alkali metals. This type of ion source is currently being used for ion rocket engines.

### 7.2.8 Electron Attachment

An electron and a neutral gas atom or molecule may become attached and form a negative ion. The probability of this process occurring depends on the energy of the electron and the nature of the gas. $O^-$, $O_2^-$, $NO_2^-$, $NO_3^-$, $OH^-$, $H^-$, $Li^-$, $CH^-$, $C^-$, and the halogens are readily formed but not $N^-$, $N_2^-$, or negative ions of the rare gases (ref. 5). In the attachment process

$$A + e + KE \rightarrow A^- + KE' + E_a \tag{7–2}$$

the energy liberated is the kinetic energy ($KE$) plus the electron affinity $E_a$ and appears in the form of radiation. The electron affinity represents the binding energy of the attached electron, a high value signifying a firmly bound electron. The electron affinity is about 2 ev for O, 1 ev for $O_2$, and 0.7 ev for H.

When an electron becomes attached to a molecule, the energy liberated may be converted to potential energy. For example, the molecule may first be excited to a higher state and then dissociate into an excited atom and a negative atomic ion.

The probability of attachment increases as the electron energy decreases, because the electron remains within the sphere of influence of the atom for a longer time. The probability of attachment depends on the chance that a quantum of energy, $E_a + KE$, will be emitted during this time.

It should be noted that electron attachment may be part of a recombination process. An electron may first attach to a neutral atom, and the resulting negative ion combines with a positive ion to form a neutral molecule or two neutral atoms.

## 7.3 Neutralization Processes

Neutralization (or recombination) is the process in which negative and positive particles come together to form neutral particles. For the purpose of analysis it is assumed that there is just one species each of negative and positive particles, and that they are distributed uniformly in space. The rate of recombination is proportional to the number of encounters between particles of unlike charge, which in turn is proportional to the number of positive particles $n_+$ and the number of negative particles $n_-$ (the encounters may be any type). Therefore, the loss in the number of charged particles $dn$ in a time $dt$ is

$$dn = -\alpha n_+ n_- \, dt \tag{7-3}$$

where $\alpha$ is the coefficient of recombination. If $n_+ = n_-$,

$$dn/dt = -\alpha n^2 \tag{7-4}$$

The coefficient of recombination varies with the gas, its pressure and temperature, the ions present, etc. Assuming these conditions constant, the equation can be integrated directly:

$$\int_{n_0}^{n} \frac{1}{n^2} \, dn = -\alpha \int_0^t dt \tag{7-5}$$

which gives

$$\alpha = \frac{1}{t}\left(\frac{1}{n} - \frac{1}{n_0}\right). \tag{7-6}$$

This result gives a means of determining the value of $\alpha$. If $n$ can be measured and $1/n$ is plotted versus $t$, the slope of the line will be equal to $\alpha$.

It is found in experiments that rate of recombinations can greatly exceed the rate of collisions between the gas particles as calculated from kinetic theory. This difference is due to the attractive forces which cause particles of opposite charges to drift together. The nature of the recombination process will depend on the temperature and density of the gas as well as the number density and type of charged particle. The first investigator to consider all of these factors was J. J. Thomson in his ion-ion three-body recombination theory.

Thomson postulated that in order for recombination to take place, the Coulomb potential energy of the particles must be greater than the mean kinetic energy (or random thermal energy) of the particles (ref. 8). That is, assuming singly ionized particles,

$$e^2/d_0 > \tfrac{3}{2}kT.$$

Other approaches to this problem give different values for this numerical coefficient depending on whether the energy of the ions is for a single ion or both ions, etc.

The radius of the sphere of active attraction is defined by

$$d_0 = \tfrac{2}{3}e^2/kT \tag{7-7}$$

where $e$ is the electronic charge, $k$ is the Boltzmann constant, and $T$ is the absolute temperature. It defines a region about the ion within which this ion and an ion of opposite sign will have the attractive Coulomb force of sufficient magnitude to change a random diffusive drift to a more directed movement toward each other.

The Thomson mechanism involves four basic steps or periods as follows (ref. 9):

(1) *Diffuse approach period:* During this period the ions diffuse to within $d_0$ of each other. It is only of significance when $r_0 > d_0$, where $r_0 = n_0^{-1/3}$ and $n_0$ is the initial number density of ions. For electron-ion recombination, the electron can be thought of as diffusing to within a sphere representing the capture cross section of the ion, where neutralization occurs. The evaluation of this cross section must be determined from quantum mechanical analysis.

(2) *Period of active attraction within the sphere of attraction:* During this period, the ions will drift toward each other under their mutual Coulomb attractions. If $d_0 \gg \lambda$, where $\lambda$ is the mean free path, then the ions will move together with their drift velocity $v = e/r^2(\mu_+ + \mu_-)$ where $r$ is the distance between the particles at any instant and $\mu_+$ and $\mu_-$ are the mobilities of the positive and negative ions, respectively. Mobility is defined by

$$\mu = \bar{v}/X \tag{7-8}$$

where $\bar{v}$ is average velocity and $X$ is the electric field strength. If $\lambda \approx d_0$, the ions will move together in a discontinuous series of a few steps. If $\lambda > d_0$, practically all of the motion within $d_0$ will be the execution of the Coulomb force orbits.

(3) *Period of orbital encounter:* Once the ions are separated by a distance of the order of $\lambda$, they will execute orbits about the common center of mass, i.e., there is a period of orbital encounter. If the kinetic energy in the orbit is greater than the minimum potential energy, the ions will separate. If the kinetic energy is less, the ion will move in an elliptical orbit until the next impact with a neutral molecule at which time the ions are either forced closer together, or further apart. In the latter case the ion may escape from $d_0$.

(4) *Period of charge transfer:* If, during the period of orbital encounter, the electron of the negative ion can tunnel through the barrier within the ion and reach the positive ion, charge transfer and neutralization result. Also, if the positive ion potential well approaches close enough to the negative ion during the orbital encounter, the potential barrier of the negative ion will be much lowered. In this event, the electron will spill over to the positive ion and will be captured into an orbit. Since the ionization energy of the positive ion is usually larger than the negative ion formation, energy, the balance of the energy will go to kinetic energy of separation of the neutralized carriers.

The four-step Thomson process, although postulated for ion-ion recombination, is also applicable, with slight modifications, to electron-ion recombination.

Recombination processes can be grouped as follows (refs. 10 and 11):

I. Ion-ion recombination
- A. Normal ion-large ion recombination
- B. Normal ion-normal ion recombination
  1. Preferential recombination
  2. Volume recombination
     - (a) High pressure process
     - (b) Intermediate pressure process
     - (c) Low pressure process
       - (1) Radiative recombination
       - (2) Mutual neutralization by charge exchange

C. Columnar recombination

D. Initial recombination

II. Electron-ion recombination

A. Preferential recombination

B. Volume recombination

1. Radiative capture
2. Dissociative capture
3. Dielectronic capture
4. Three-body capture

C. Wall recombination

D. Beam recombination

Large ions refer to very small solid particles, such as dust, smoke, etc., which are ionized. This process does not seem to have much importance in gas dynamics and will not be discussed further.

The ion-ion processes which are the most important are preferential and volume recombination. Preferential recombination occurs at very high pressures, and volume recombination occurs between particles isotropically and randomly distributed. Columnar recombination has been observed in alpha-particle tracks and is due to the anisotropic distribution of ions. Initial recombination occurs with X-ray or $\beta$-particle ionization and is caused by rapid electron attachment which creates ions in close proximity. Columnar and initial recombinations will not be discussed further in this chapter.

Of the electron-ion recombination processes, volume and wall recombinations are the most important. Preferential electron-ion recombination occurs only at extremely high pressures and therefore will not be discussed further.

### 7.3.1 Ion-Ion Preferential Recombination

This process occurs in electron attaching gases at high pressures for which $d_0 \gg r_0 \gg \lambda$ (ref. 10). Under these conditions the ions drift toward each other as a result of their Coulomb attraction. The drift is retarded by collisions with gas atoms, and therefore the relative speed between the ions will be small compared with the thermal speed. From simple theoretical considerations, the recombination coefficient is shown to be given by (ref. 5)

$$\alpha = 4\pi e(\mu_+ + \mu_-). \tag{7–9}$$

The mobility varies inversely with the pressure (ref. 5), and therefore the recombination coefficient also varies inversely with the pressure.

This model and the resulting expression for $\alpha$ were derived originally by Langevin, and this expression is known as the Langevin law. It is applicable at very high pressures of the order of 100 atm, although it gives good results down to about 2 atm.

It follows that under the conditions for which this theory is applicable, ions of the opposite sign always drift together. Therefore, if the ionization was initially anisotropic, the anisotropy would remain, yielding $\alpha$ constant with time, hence the name "preferential recombination" (ref. 10). All of the observations at 1 atm and below show the opposite behavior.

### 7.3.2 Ion-Ion Recombination

#### *7.3.2.1 High Pressures*

This process is similar to preferential recombination except that the average distance between the ions is greater than the radius of the sphere of active attraction, i.e., $r_0 \gg d_0 \gg \lambda$. The chance of charge transfer once the ion is within $d_0$ will be practically unity, since it is probable that a collision with a neutral particle will occur in this region. Thus, the rate of recombination should depend mainly on the rate of diffusion of the ions to within a distance $d_0$ plus the time required for neutralization within $d_0$. An expression for the recombination coefficient has been derived as (ref. 10)

$$\alpha = 4\pi ef(\mu_+ + \mu_-) \tag{7–10}$$

which is the same expression as for preferential recombination multiplied by $f(\simeq 1)$. It can be concluded, therefore, that irrespective of whether the ions must first diffuse together as in volume recombination, or are always drawn together as in preferential recombination, when $d_0 \gg \lambda$, the expression for $\alpha$ is of the same form.

### *7.3.2.2 Intermediate Pressures*

This process occurs in the pressure range of approximately $10^3$ to $10^{-2}$ mm Hg with $r_0 \gg d_0 \gtrless \lambda$. The four-step Thomson process occurs, and the recombination coefficient can be predicted from the successful theory of three-body recombination developed by Thomson and later modified by others.

The recombination coefficient can be written in the form (ref. 10)

$$\alpha = \pi d_0{}^2 \varepsilon \sqrt{v_+{}^2 + v_-{}^2} \tag{7-11}$$

where $v_+$ and $v_-$ are the average thermal velocities of the positive and negative ions, respectively, and $\varepsilon$ is the probability that one of the ions will collide with a neutral particle within $d_0$ so that recombination can occur. Thomson derived the value of $\varepsilon$

$$\varepsilon = w_+ + w_- - w_+ w_-, \tag{7-12}$$

where $w_+$ and $w_-$ are chances of encounter within $d_0$ for the positive and negative ions respectively. It can usually be assumed that $w_+ = w_-$ and $v_+ = v_-$, so the recombination coefficient becomes

$$\alpha = \sqrt{2}\, \pi d_0{}^2 v(2w - w^2). \tag{7-13}$$

Since $d_0{}^2$ is proportional to $T^{-2}$ and $v$ is proportional to $T^{1/2}$, $\alpha$ varies as $T^{-3/2}$ at constant pressure.† At high pressures, however, $w$ approaches unity. Thus, the above relation indicates that, at constant temperature, $\alpha$ should approach a constant value. Since it has been shown that at high pressures the recombination

---

† At constant temperature, $\alpha$ increases with pressure, exhibiting just the opposite behavior as for high pressures.

coefficient varies inversely with pressure, it can be concluded that the Thompson process is not applicable for the high pressure region.

### *7.3.2.3 Low Pressures*

Low pressure ion-ion volume recombination occurs at pressures below about $10^{-3}$ mm Hg with $r_0 \gg d_0$ and $\lambda \gg d_0$. In this case the value of $\varepsilon$ becomes very small and Eq. (7–12) cannot be used. The recombination coefficient, however, can be estimated from (ref. 6)

$$\alpha = \pi\sigma^2 \sqrt{v_+{}^2 + v_-{}^2} \tag{7–14}$$

where $\sigma$ is defined as the distance between the oppositely charged ions at which the chance of charge transfer approaches unity.

Very little experimental data have been obtained for recombination at low pressures. Measurements are difficult to make at low pressures as a result of ambipolar diffusion and wall recombination. [In a plasma in which there are ions of both signs, the faster ions, generally electrons, tend to diffuse out of the plasma, leaving an excess of positive charge whose space charge field then retards the electrons and accelerates the positive ions, thus linking their diffusion. This process is known as ambipolar diffusion (ref. 9).]

At very low pressures, it is very unlikely that three-body recombination will occur. The energy involved in the recombination, however, must be removed by some mechanism. Two processes which are likely to occur at low pressures are radiative recombination and mutual neutralization by charge exchange (ref. 11).

The radiative recombination reaction is

$$A^- + B^+ + KE \rightarrow AB + h\nu + KE'. \tag{7–15}$$

If all of the energy involved in the process is radiated, i.e., $KE = KE'$, $h\nu$ must be equal to the ionization energy of B plus the dissociation energy of AB minus the electron affinity of A.

The reaction for mutual neutralization by charge exchange is

$$A^- + B^+ + KE \rightarrow A^* + B^* + KE', \tag{7-16a}$$

$$A^* + B^* + KE' \rightarrow A + B + KE' + h\nu, \tag{7-16b}$$

where A* and B* refer to possible excited states of A and B. What the above reaction means is that in the neutralization process, either A or B or both may become excited. If both atoms eventually return to their normal states, and all of the energy involved in the process is radiated ($KE = KE'$), $h\nu$ is equal to the ionization energy of B plus the electron affinity of A.

It should be noted that either radiative recombination or mutual neutralization can also occur at higher pressures. However, they most likely will not be of significance because three-body recombination is much more probable at high pressures.

### 7.3.3 Electron-Ion Volume Recombination

#### *7.3.3.1 Radiative Capture*

This capture occurs in a gas in which the atoms or molecules do not dissociate upon recombination with an electron. The captured electrons, which have a continuous energy distribution, are located in a previously unoccupied level of the atom. The reaction can be written as

$$A^+ + e^- + \tfrac{1}{2} m_e v^2 \rightarrow A + h\nu \tag{7-17}$$

where

$$h\nu = E_I + \tfrac{1}{2} m_e v^2. \tag{7-18}$$

The coefficient of recombination can be found from (ref. 10)

$$\alpha = \pi\sigma^2 v \tag{7-19}$$

where $\pi\sigma^2$ is the atomic cross section for radiative capture and $v$ is the mean thermal velocity of the electrons. The cross section can be calculated from quantum mechanics for simple hydrogen-like atoms. For more complicated atoms or molecules, the cross section must be determined experimentally. There are also approximate formulas from which $\alpha$ can be estimated (ref. 5).

*7.3.3.2 Dissociative Capture*

In a polyatomic gas, the molecules may dissociate after recombination if the energy involved in the process is large enough. The reaction is

$$A_2^+ + e^- + \tfrac{1}{2} m_e v^2 \rightarrow A + A^* + \tfrac{1}{2} m_A v^2 \rightarrow 2A + \tfrac{1}{2} m_A v^2 + h\nu \qquad (7\text{–}20)$$

where A* refers to an excited state of A. The difference between the energy of neutralization and the dissociation energy of $A_2$ goes into the kinetic energy of the dissociated molecule and possibly also into the energy of excitation of one or both of the atoms. The excited atoms will radiate energy when they return to their normal states. A theory has been developed for this process by Loeb (ref. 10).

It is found experimentally that all molecular ions may not recombine dissociatively, even if the energy balance is in favor of this process. However, dissociative recombination is commonly observed. For example, it is the main recombination process in ionized nitrogen gas at low temperatures (ref. 12).

*7.3.3.3 Dielectronic Capture*

This process is the inverse of an ionization process called autoionization, in which two electrons in an atom which have been raised to excited states possess a total energy greater than the ionization energy of the atom. When one electron returns to its ground state, the other leaves the atom. In the inverse process, when the electron is captured by the atom, the surplus energy (which is equal to ionization energy minus the excitation energy of the captured electron) is transferred to a second electron in the atom thereby exciting it. These excited electrons may later radiate their energy and return to their normal states. The reaction is

$$A^+ + e^- + \tfrac{1}{2} m_e v^2 \rightarrow A^* \rightarrow A + h\nu. \qquad (7\text{–}21)$$

This process is not very common, and the recombination coefficient has not been determined experimentally. However, a theory has been developed from which the recombination coefficient can be estimated (ref. 11).

#### *7.3.3.4 Three-Body Capture*

This process is similar to ion-ion three-body recombination. The third body could be an atom, molecule, ion, or another electron. The Thomson three-body recombination theory may be extended to apply to electrons if some modifications are made (ref. 11).

### 7.3.4 Electron-Ion Wall Recombination

This process generally takes place in a confined plasma at low pressures. In many cases, the recombination at the wall takes place more rapidly than the diffusion of charged particles to the wall (ref. 9). Thus, the loss of charged particles from the gas is governed by ambipolar diffusion.

The recombination coefficient at the wall will depend on the surface material. The surface may either enhance recombination (catalytic wall) or inhibit recombination (poisoned wall).

This process also occurs at high pressures, although it may not be of great importance compared to the rate of recombination taking place away from the wall. Wall recombination undoubtedly takes place on the surface of a body traveling at hypersonic speeds. In this case, wall recombination may be of importance, since it may affect the heat transfer to the wall.

### 7.3.5 Electron-Ion Beam Recombination

A recombination process which bears special mention is recombination of a beam of positive particles with a beam of negative particles. Until recently, this problem was of academic interest only, and was investigated very little. Recently this problem has become important in the neutralization of positive ion space charge in an ion rocket engine exhaust. It is necessary to neutralize this space charge with beams of electrons, and it would be desirable to effect recombination of the beams of ions and electrons.

Barnes (ref. 13) describes experiments in which alpha particles traveled through a cloud of electrons moving in the same direction. As would be expected, the percentage of alpha particles which combined with at least one electron was high when the velocities of the alpha particles and electrons were equal. However, the percentage was found to be just as high when the electron velocity equaled the alpha particle velocity plus or minus the electron velocity in the various quantum states of the helium atom. The time of recombination was found to be less than $3 \times 10^{-10}$ sec, and the percentage of alpha particles which captured an electron in this time was found to be as high as 90%. However, the electrons greatly outnumbered the alpha particles (a single alpha particle traveled through an electron cloud of density $\approx 10^7$ per cc), while in the ion rocket engine exhaust, the number of ions and electrons will be approximately equal. Therefore, it seems likely that the percentage of capture in the rocket exhaust within a reasonably short distance will be rather low.

### 7.3.6 Experimental Values of Recombination Coefficients

Although gas purity, reliability of experimental techniques, and similar considerations cast serious doubts on the value of

TABLE 7–1

DATA FOR ION-ION RECOMBINATION

| Gas | Ion concentration per $cm^3$ | Temp. (°C) | Pressure (mm Hg) | Recombination coefficient |
|---|---|---|---|---|
| $O_2$ | $2 \times 10^6$ | 25 | 760 | $2.08 \times 10^{-6}$ |
| $O_2$ | $10^5$ | 25 | 760 | $2.5 \times 10^{-6}$ |
| Air | $1.5 \times 10^6$ | 20 | 760 | $2.65 \times 10^{-6}$ |
| Air | $4.4 \times 10^6$ | 20 | 760 | $2.3 \times 10^{-6}$ |

most experimentally determined recombination rate coefficients, Loeb (ref. 10) has studied data on ion-ion recombination in oxygen and air. His representative values are given in Table 7–1. Figures 7–3 to 7–6 show the trends commonly accepted for the variation

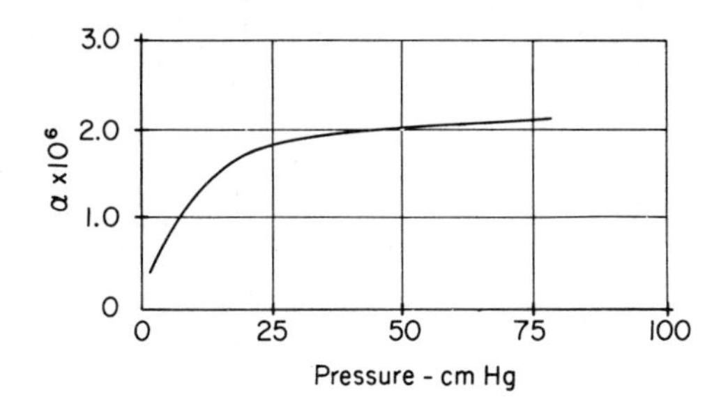


FIG. 7–3. Recombination coefficient in oxygen versus pressure.

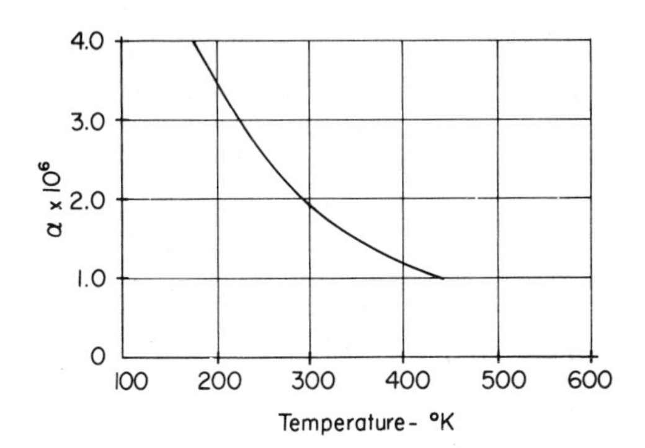


FIG. 7–4. Recombination coefficient in oxygen versus temperature.

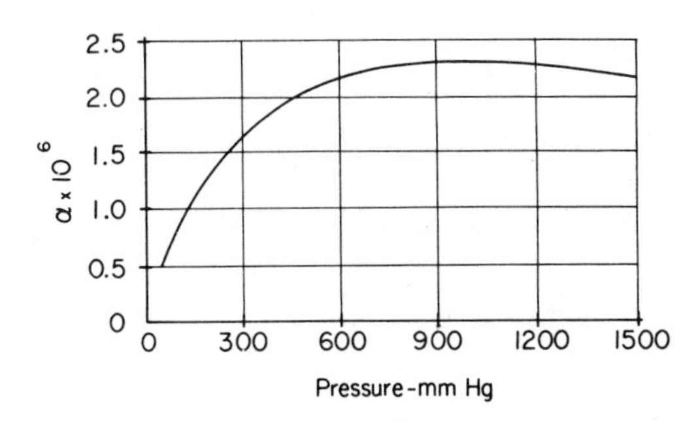


FIG. 7–5. Recombination coefficient in air versus pressure.

of the ion-ion recombination rate with both temperature and pressure (in oxygen and air). A very significant feature of Fig. 7–6 is that the slope is positive at low pressures and negative at higher

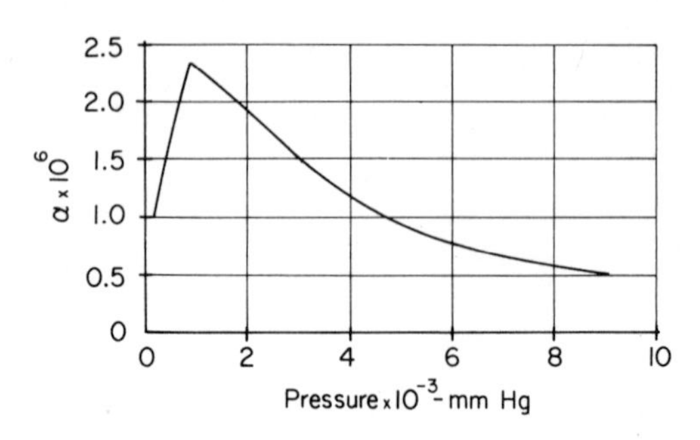


FIG. 7–6. Recombination coefficient in air versus pressure.

pressures. This seems to verify the existence of different recombination mechanisms at different pressures, seriously complicating theoretical approaches to the problem.

## 7.4 List of Symbols

Latin symbols:

A — typical atom
$d_0$ — radius of sphere of attraction
$e$ — electron charge
$E$ — energy
$E_a$ — electron affinity
$f$ — constant (near unity)
$h$ — Planck's constant
$I$ — intensity
$k$ — Boltzmann's constant
$KE$ — kinetic energy
$m$ — mass
$n$ — number density

$r$ — distance between particles
$S$ — number of ion pairs
$t$ — time
$T$ — absolute temperature
$v$ — velocity
$w$ — probability of encounter
$x$ — distance
$X$ — electric field strength

Greek symbols:

$\alpha$ — coefficient of recombination
$\varepsilon$ — collision probability
$\lambda$ — mean free path
$\mu$ — mobility
$\nu$ — frequency
$\sigma$ — charge transfer coefficient

Subscripts:

A — related to neutral atoms
$e$ — related to electrons
I — ionization
0 — reference value
$+$ — related to positive ions

Superscripts:

$*$ — excited state
— — average value

## References

1. Cobine, J. D., *Gaseous Conductors.* Dover Publications, New York, 1958.
2. Compton, K. T., and I. Langmuir, "Electrical Discharges in Gases. I. Survey of Fundamental Processes." *Reviews of Modern Physics*, Vol. 2, No. 2, April, 1930, pp. 123–242.
3. Maxfield, F. A., and R. R. Benedict, *Theory of Gaseous Conductors and Electronics.* McGraw-Hill, New York, 1941.

4. Francis, V. J., and H. G. Jenkins, "Electrical Discharges in Gases and their Applications, I." *Reports on Progress in Physics*, Vol. 7, 1940, pp. 230–302.
5. von Engel, A., *Ionized Gases*. Oxford Univ. Press, London and New York, 1955.
6. Llewellyn-Jones, F. *Ionization and Breakdown in Gases*. Methuen, London, 1957.
7. Datz, S., and E. H. Taylor, "Ionization on Platinum and Tungsten Surfaces.I. The Alkali Metals." *Journal of Chemical Physics*, Vol. 25, No. 3, September, 1956, pp. 389–394.
8. Thomson, J. J., and G. P. Thomson, *Conduction of Electricity Through Gases*, Vol. I, 3rd edition. Cambridge Univ. Press, London and New York, 1928.
9. Loeb, L. B., *Basic Processes of Gaseous Electronics*. Univ. of California Press, Berkeley, California, 1955.
10. Loeb, L. B., "The Recombination of Ions," in *Handbuch der Physik* (edited by S. Flügge). Springer-Verlag, Berlin, 1956.
11. Massey, H. S. W., and E. H. S. Burhop, *Electronic and Ionic Impact Phenomena*. Oxford Univ. Press, London and New York, 1952.
12. Bialecke, E. P., and A. A. Dougal, "Pressure and Temperature Variation of the Electron-Ion Recombination Coefficient in Nitrogen." *Journal of Geophysical Research*, Vol. 63, No. 3, September, 1958, pp. 539–546.
13. Barnes, A. H., "The Capture of Electrons by Alpha Particles." *Physical Review*, Vol. 35, No. 3, February 1, 1930, pp. 217–228.

# AUTHOR INDEX

# SUBJECT INDEX